BOWLAND BETH

BOWLAND BETH

The Life of an English Hen Harrier

DAVID COBHAM

ILLUSTRATED BY DAN POWELL

William Collins
An imprint of HarperCollins*Publishers*
1 London Bridge Street
London SE1 9GF

WilliamCollinsBooks.com

First published in the United Kingdom by William Collins in 2017

22 21 20 19 18 17
11 10 9 8 7 6 5 4 3 2 1

Illustrations by Dan Powell

A catalogue record for this book is available from the British Library.

ISBN 978-0-00-825189-5

Designed and typeset in Baskerville by This Side
Printed and bound by CPI Group (UK) Ltd, Croydon, CR0 4YY

MIX
Paper from
responsible sources
FSC C007454

This book is produced from independently certified FSC paper to ensure responsible forest management.

For more information visit: www.harpercollins.co.uk/green

For Stephen Murphy,
who guards the flame that
keeps alive the future of our
English hen harriers.

CONTENTS

PROLOGUE

Bowland Beth, named after the Forest of Bowland where she was bred, was an exceptional hen harrier. Some harriers, when near to fledging, stand head and shoulders above the rest of the brood. They are bigger, their plumage is glossier and richer in colour, their eyes are brighter and they fledge ahead of their siblings. Beth was one such.

Following the style of Henry Williamson's *Tarka the Otter* and Fred Bodsworth's *Last of the Curlews,* I have dramatised Beth's short life between 2011 and 2012, trying to enter her world to show what being a hen harrier today is like. I have immersed myself not only in the day-to-day regimen of her life – the hours of hunting, bathing, keeping her plumage in order and roosting – but have also attempted to express the fear of living in an environment managed to provide packs and packs of driven grouse for a few wealthy people to shoot for sport.

I hope that by dramatising Bowland Beth's life I can rally another group of conservationists to join other groups already working hard to challenge those who are determined to illegally exterminate English hen harriers in the interests of driven grouse shooting.

Nearly fifty years ago I made a film of *Tarka the Otter,* and when it was finished we held a special press show for children, mostly teenagers. Afterwards I asked the audience to come forward and ask questions. Mostly they wanted to know how we made the bubbles and blood in the river. Finally, I was left with one teenaged girl standing before me. When I asked her whether she had liked the film, she said: 'The otter died. It made me cry.' To her astonishment

I replied: 'That's wonderful. You felt something.'

I'm often told that my film of Tarka played an important role in getting otter hunting banned in England and Wales in 1980. Can this dramatised version of Beth's life awaken similar emotions in people's hearts and compel them to stand up and deplore in the strongest terms the illegal persecution of the few remaining breeding hen harriers in England? I hope so.

I have been able to dramatise Beth's life because she was fitted with a satellite tag, and the moment she fledged she was tracked as she sped off to a grouse moor at Nidderdale thirty miles away. A select few watched with delight and alarm as she foraged over a wider and wider area. She survived the winter, and in late spring her hormones kicked in and she started searching high and low for a mate. Her journeys up into the wilds of Scotland made her a national celebrity. On one journey north she covered 125 miles in just eight and a half hours, but without fail she returned to where she had been fledged in the Forest of Bowland.

I knew about the Forest of Bowland because my wife Liza had visited it whenever she could when she was performing in plays at the Grand Theatre in Blackpool. Each time she came back she was bubbling over at the number of harriers she had seen and how kind Stephen Murphy, Natural England's hen harrier project officer in Bowland, had been in showing her round.

Now, on 24 May 2012, I was going to see Bowland myself. I was researching for a book I was writing on the

state of our birds of prey in Britain today and my companion was Eddie Anderson, whom I had known for nearly fifty years. When I first met him he was a gamekeeper. Eventually, after seven years, he gave up gamekeeping and forged a fine career making programmes for Anglia TV and BBC East.

We travelled by train from Norfolk. At Clitheroe, our destination, we were met by the late Mick Carroll, one of Stephen Murphy's most dependable hen harrier watchers. He was going to drive us around Bowland, so we got into his car and he whisked us off to The Hark to Bounty pub in Slaidburn, where we were booked in for two nights.

We had a brief meeting with Stephen Murphy after breakfast. He promised to show us around the following day and then handed us over to Bill Hesketh and Bill Murphy, Stephen's local, highly regarded harrier watchers. The rain had cleared, and the fells and moorland sparkled like a newly minted coin.

The rolling hills were dominated by a patchwork of green sprouting heather set against lower areas of grass split up into tiny fields by fences and stone walls, and I could quite understand why the place had been named an Area of Outstanding Natural Beauty, a European designation that recognises the importance of its heather moorland and blanket bog as a special habitat for upland birds.

The wild, desolate landscape is an excellent location for bird watching. Curlew, golden plover, oystercatchers and snipe are all found here, and migrating dotterel pass through in spring and autumn. The summer population

of breeding curlew is one of the most buoyant in England, and short-eared owls can regularly be seen hunting in daylight for their favourite prey, short-tailed field voles.

Three important birds of prey breed in the Forest of Bowland: the merlin, our smallest bird of prey and not much bigger than a mistle thrush, the much larger peregrine falcon, which, with its 200 mph scything stoop, has been described as the most successful bird in the world, and the hen harrier. The hen harrier is the symbol of the Forest of Bowland and used to be seen regularly as it floated low over moorland hunting for small birds and field voles.

CHAPTER ONE

March 2011. *Late in the afternoon at the beginning of March, a dark chocolate-brown female hen harrier wheeled over the Forest of Bowland. The low sun enhanced her brilliant orange eyes, deep set in her owl-like face. She was looking down on the moors, still in the grip of last night's frost. Lines of grouse butts, positioned at the end of each drive, threw long shadows across the heather. A quick flick of a tail feather and a stronger down-beat on one wing brought her facing north. Below, she could pick out Ward's Stone, the highest point in the Forest. Beyond, in the far distance, her eyes, eight times sharper than ours, picked out the Wig Stones thirty miles to the north. She was hungry. Her only kill had been a meadow pipit, not much of a meal. Where were the other harriers hunting?*

She turned, flying across the Roman road towards the source of the Whitendale River. At once she spotted two harriers hunting the boggy ground draining into the river, so she spread her tail to check her speed, half closed her wings and in a series of zig-zags, the air whispering through her primaries, dropped down to join them.

Selecting an area to hunt over, she flew downwind, letting the wind do all the work, her wings held in a 'V' above her back. She scanned the hoarfrost-rimed rushes and Molinia *grasses passing slowly beneath her for any sign of movement, then turned at a dry stone wall and flew upwind. Flap, flap, flap, glide. She quartered up and down the marshy area like a person searching for something they had lost, but she was hungry and needed to make a kill.*

She saw little piles of cut grasses piled on the banks of a rivulet – signs of water voles, snug and warm in their burrows. She flapped her wings, spread her tail and slowed down. Ahead was a muddy patch that was untouched by the overnight frost. The

cryptic plumage of a common snipe blended with the zig-zag pattern of the short grasses in the background, and it was only the sudden movement as the snipe plunged its sensitive bill into the mud that gave it away. The harrier suddenly flapped her wings once, stretched her legs down in line with her targeting eye and plunged down, her needle-sharp talons crushing the life out of the snipe.

She carried her kill away to a dense clump of rushes where she could pluck it and feed undisturbed. Ten minutes later, hunger satisfied, she joined the other harriers that were making their way back to the communal roost site at Brackenholes Clough.

The story of Bowland Beth is indelibly linked to Stephen Murphy, who leads the Natural England mission to safeguard the few remaining hen harriers in Bowland. I have known Stephen since 2006, when he radio-tagged a pair of marsh harrier chicks at the Hawk and Owl Trust reserve at Sculthorpe Moor in Norfolk. With the help of my original interviews with him in 2012 – and many recorded over the years 2014 to 2016 – I have been able to piece together the events that ushered Bowland Beth into the world.

I am looking at a photograph of him as I write. He has close-cropped hair, fine-chiselled features and is grinning at the camera because he is holding a month-old hen harrier chick on his knee. He radiates a natural enthusiasm for hen harriers, infecting everyone he meets with his love for these birds.

'My dream job at Bowland, which I started in 2002, was seen, by some, as a poisoned chalice,' he says. 'Bowland is

a place I love dearly. It had once been the stronghold of breeding pairs of hen harriers in England – twenty-plus pairs in the 1980s. From then on, the productivity of the birds yo-yo'd up and down every two or three years. Now, in 2012, we're down to one breeding pair.'

A survey of hen harriers in the British Isles was carried out in 2010. There were then 630 pairs, with the great majority – five hundred-plus pairs – in Scotland, small populations in Wales and Northern Ireland of just under sixty pairs, and a diminishing population of thirty pairs on the Isle of Man. In England the breeding population was a mere twelve pairs, although conservationists tell us that there is suitable moorland habitat for three hundred-plus pairs.

March 2011, *Norfolk. The cold darkness before sunrise after a sharp overnight frost. The three-year-old male hen harrier stood up from his roosting place, a soft couch in an area of sphagnum bog surrounded by rushes, heather and dwarf silver birch. Deep within his body chemical changes were taking place, urging him to fly north and seek a mate, forces that would shape everything that would happen in the future. He scanned the roost and counted the dark shapes – four other harriers nearby. It was common land and had been used by harriers as a safe roosting place for centuries.*

The harrier stepped forward and cast up a pellet, an oblong mass of fur, feather and bones. He had arrived two days earlier, having overwintered in France, and was now in peak condition. The first light of the rising sun showed a very handsome bird: bright orange eyes, the cere above his sharply curved beak bright

yellow, his body – apart from his scapulars and mantle, which were a dusky brown – covered in a silver-grey plumage, black primaries, grey tail and long yellow legs with sharp black talons. He took off, and as he circled higher and higher he saw beneath him the sugar beet fields, where yesterday he had hunted for finches, pipits and migrant thrushes. A few pink-footed geese, stragglers, were flighting in to feed on the sugar beet tops. Further on he skirted the Wash, a vast mud flat separating Norfolk from Lincolnshire. He saw huge flocks of wading birds taking to the air and flying inland as the rising tide covered the glistening mud.

Deep within his brain, information stored from his first view of the nest in which he was hatched directed the sturdy pectoral muscles in his chest, forcing his spread wings down, driving him forward. At the end of each stroke other muscles retracted the wing, ready for the next driving forward stroke. He was heading two hundred miles north-west to the Forest of Bowland, where he had been born three years ago.

Halfway through the Mesolithic Age, ten thousand years ago, the hen harrier would have been a common sight in England. Then, there were millions of acres of heaths, moors, mountains and barren lands. These, together with undrained bogs and marshes, would have provided huge areas of suitable hunting and nesting habitat for harriers to enjoy. The late Derek Yalden and Umberto Albarella found just such a landscape in the Bialowieza National Park on the Polish–Byelorussian border, and in their historical reconstruction they described what bird life would have been like in Britain at that time. Taking the

average densities in areas studied by experts in Britain and combining them with the fact that there would have been no persecution or interference, Yalden and Albarella were able to calculate a Mesolithic hen harrier population of 2,803 pairs.

Mesolithic man, a hunter-gatherer, made little impact on the landscape. It was Neolithic man, who arrived in Britain five thousand years ago, who began clearing forest and scrub to make fields. He planted grain crops and his livestock grazed the grasslands on the chalk areas. The sophistication of farming increased during the Bronze Age, when the importance of soils and a good supply of water became understood. The village, in its simplest form, was a cluster of dwellings surrounded by three common fields, fertile and easy to farm, with ox-drawn ploughs being used to strip-plough the land. One of the fields would be planted with wheat or rye for bread, another with grains for beverages and the third would lie fallow. After the Norman Conquest churches were built in each village, the population quadrupled, and farming and clearance of the scrub and forest accelerated. But the yields of grain were poor, barely sufficient to feed the population.

The Black Death ravaged the country in the fourteenth century, halving the population. Cultivated land was abandoned and farmers turned to raising sheep. Wool made some farmers a fortune and sheep, a walking dung cart, improved the fertility of the fields.

The Tudor Vermin Acts, initiated in 1532 and strengthened in 1566, listed the birds and mammals

considered to be vermin. The list was extensive, ranging from stoats and weasels through bullfinches and kingfishers to ospreys and so-called ringtails, as the hen harrier was known. Not surprisingly, it was on the list; it has always been a controversial bird, as its name implies, and must have been a familiar sight as it hunted around farmers' smallholdings, snatching up chicks and ducklings. Churchwardens were responsible for payments to those who destroyed vermin, although it appears that few payments were made on this account.

The Enclosures Acts of the eighteenth and nineteenth centuries meant that the days of the hen harrier were numbered, even though they remained common enough for the poet John Clare to watch them float languidly over the fens near his home in Northamptonshire. 'There is a large blue (hawk) almost as big as a goose,' he wrote, '[that] fly in a swopping manner not much unlike the flye of a heron you may see an odd one often in the spring swimming close to the green corn and ranging over an whole field for hours together – it hunts leverets, partridges and pheasants.'

In his poem 'The Shepherd's Calendar', John Clare paints a vivid picture of a hunting hen harrier.

A hugh [huge] blue bird will often swim
Along the wheat when skys grow dim
Wi clouds – slow as the gales of spring
In motion wi dark shadowed wing
Beneath the coming storm it sails.

One of the hen harrier's main habitats was heathland, created when woodland was cut down. This was treated as common land, and among its uses was grazing for sheep, furze for heating – and also for burning heretics – and ling for fuel and low-quality thatch. But heathlands became increasingly threatened. In *The Natural History and Antiquities of Selborne*, published in 1789, the naturalist Gilbert White raged at the senseless, illegal burning of heathland:

> To burn on any waste, between Candlemass and Midsummer, any grig, ling, heath and furze, goss or fern, is punishable with whipping and confinement in the house of correction.

He takes a deep breath.

> The plea for these burnings is, that, when the old coat of heath, etc, is consumed, young will sprout up, and afford much tender brouze for cattle; but, where there is large old furze, the fire, following the roots, consumes the very ground; so that for hundreds of acres nothing is to be seen but smother and desolation, the whole circuit around looking like the cinders of a volcano; and, the soil being quite exhausted, no traces of vegetation are to be found for years.

In some heathlands there was no grass to bind the sand, and the land often resembled the Sahel in Africa. When there was a big 'blow', as in the seventeenth century in the Norfolk village of Downham, settlements could be almost buried in sand and rivers blocked. A farmer, asked where he lived, replied, 'Sometimes in Norfolk and sometimes in

Suffolk; it depends which way the wind blows.'

Gradually, much of the hen harrier's favourite habitats – the heaths and fens – vanished as the Enclosures Acts took hold. But the face of Britain was completely altered by the Agricultural Revolution, particularly the four-course system promoted by Charles 'Turnip' Townshend. This meant that cattle, fed on turnips and kept during the winter, didn't have to be slaughtered in the autumn. As a direct result of this innovation, more hedges were planted to contain the cattle.

A great swathe of land from north Yorkshire to Southampton was enclosed to become highly productive farmland. The landscape as we know it today became

established and hen harriers ceased to be a commonplace sight; their numbers greatly reduced, they retreated to the heather moorlands of Lancashire and north Yorkshire.

Here they found a diverse habitat. There were swathes of heather up to four feet tall – excellent cover in which to nest – whose dark understorey was rich in insects, attracting small birds like stonechats, a good, staple prey item for hen harriers. Other areas on the moorland were grassy. Hunting over these areas, the harriers caught meadow pipits and skylarks, as well as short-tailed field voles that made their burrows in these rough meadows. For more substantial fare, young rabbits and leverets could be pounced on before they were aware of the danger.

Here too were wet areas, with spongy sphagnum bogs dominating the lower reaches of the heather moorland.

Along the rivulets draining into these bogs – and in the bogs themselves – harriers could catch water voles, common snipe and curlew.

And lastly, as we shall shortly find out, there were red grouse.

CHAPTER TWO

The sun was blood red as it broke the horizon and lit the communal roost where the female hen harrier had spent the night. She watched the other harriers as they left to go foraging for food out on the moor. She didn't join them, for she had felt a quickening in her body, an urge to move to Mallowdale Pike, a rocky crag from where she had fledged nine months ago. After preening – getting her feathers into flying order – she lifted off from the roost and soared up over the fell.

Soon she was able to see the familiar mosaic pattern, the result of annual heather burning. She twitched her tail feathers on one side and completed her refamiliarisation with the grouse moor beneath her. But then she suddenly sheered away. There was something wrong. Smoke was billowing up and she could see men beating at the flames beneath her. Why were they destroying her birth place?

The first mention of the hen harrier was in 1544 by Dr William Turner in his *Avium praecipuarum*, the first bird book to be printed, during a comparison of his field observations with those made by Aristotle and Pliny.

> The Rubetarius I think to be that Hawk which English people name Hen Harroer. Further it gets its name among our countryman, from butchering their fowls. It exceeds the Palumbarius in size and is in colour ashen. It suddenly strikes birds when sitting in the fields upon the ground, as well as fowls in towns and villages. Baulked of its prey it steals off silently, nor does it ever make a second swoop. It flies along the ground the most of all. The Subbuteo I think to be that hawk

> which Englishmen call Ringtail from the ring of white that reaches round the tail. In colour it is midway from fulvous to black; it is a little smaller than the Buteo, but much more active. It catches prey in the same manner as the bird above.

Turner treated both birds as if they were different species. All ashen-coloured harriers, the male birds, were known as hen harriers, while the brown 'subbuteos', with white around the tail, were called ringtails.

This dimorphism of the harrier was a riddle that was not solved for another three hundred years. Thomas Bewick in his *History of British Birds*, published in 1797, added a cautionary footnote to his description of the hen harrier: 'It has been supposed that this and the following are male and female; but the repeated instances of hen harriers of both sexes having been seen leave it beyond all doubt that they constitute two distinct species.' At around this time John Latham made an extremely sensible suggestion that would cut through all the hot air that had been generated – take some chicks from the nest and keep them in captivity for three or four years to confirm the change in plumage.

George Montagu now applied himself to solving the 'harrier problem'. In 1807 he wrote a paper that stated two vital facts. The bird familiarly called a ringtail and given the scientific name *pygargus* was in fact the female hen harrier, *cyaneus*. In his meticulous way he described how he had taken a nest of three hen harrier chicks and

reared them in captivity – exactly what John Latham had suggested. One died, but eventually the remaining two moulted into adult plumage, a grey male and a female – a ringtail. Having done this, George Montagu was able to properly describe them as two different species, the silver-grey male hen harrier and the smaller, more graceful, silver-grey Montagu's harrier. The hen harrier was given the Latin name *Circus cyaneus*, while the Montagu's harrier was now known as *Circus pygargus*. Their females, sombre brown with a conspicuous white rump, were known as ringtails.

Later that century, J. H. Gurney, not to be outdone, reared three hen harrier nestlings taken from a marsh near Ranworth decoy in Norfolk. When fully fledged, they all displayed the rich chocolate colour of their immature plumage. On moulting, two out of the three proved to be males. They survived for five years and one is now preserved in the Castle Museum in Norwich.

I was recently allowed to study four skins of female and male hen harriers and Montagu's harriers on loan from the Castle Museum. The female hen harrier was enormous, measuring 28 inches from beak to tip of tail. Maybe it was a Scandinavian bird, which are bigger. The skin of the male Montagu's harrier demonstrated perfectly one of the key identifying features – the primaries projected beyond the tip of the tail.

It is time now to examine the fully adult bird in more detail. The female hen harrier and the immature male are rather dull compared with the adult male. The female's head and nape are light brown with dark streaks and

the back is dark brown. The secondaries are barred and the primaries are dark. The rump is white and the tail feathers have three narrow transverse dark bars and a much broader bar at the tip. The underside is pale brown with longitudinal dark brown streaking. The female is considerably larger than the male.

By comparison the male is a spectacularly handsome bird. The beak is black and the cere yellow, bristles cover the area between the cere and the eyes, which are a clear yellow, and there is a distinct owl-like, facial ruff edged with short, very distinct feathers. The head, nape, upper back and upper wing coverts of a fully adult, five-year-old bird are silver-grey. The rump is white, contrasting with the tail feathers, which are light grey with dark transverse bars and have white tips, apart from the two central tail feathers, which are plain grey and unmarked. The first five outer primaries are black, the underside is grey, becoming lighter towards the vent, and the legs are yellow and long.

CHAPTER THREE

The male hen harrier, on his way back to Bowland, had overnighted in that part of the Peak District known as the Dark Peat. As he took off from his roost where he'd been safe under cover, hidden from sight by a few stunted trees and warm in the comfort of the sphagnum bog, he noticed the almost complete absence of other birds of prey – no harriers, peregrines or goshawks, just the occasional merlin or kestrel. It was not a good place to be a bird of prey. Like all hen harriers he had a distrust of man. The sight of a line of men managing a heather burn made him jink to one side to avoid the billowing flames and smoke.

Much of the heather moorland is managed for the benefit of those who shoot red grouse. The birds are driven towards the 'guns' waiting in a line of butts set up across the moor during the grouse-shooting season, which starts on 12 August and ends on 10 December. There are four main grouse moors in Bowland – United Utilities (formerly North West Water Authority), Abbeystead, Bleasdale and Clapham estates. There are other smaller moors, all of which are important players in their own way.

The United Utilities estate is not a true moor, nor is it managed to the extent of the others. There is some driven grouse shooting but the majority of the grouse killed are shot by a single gun walking the moor with setter and pointer dogs ranging ahead. If they find grouse, the pointer holds its point on the crouching birds. The gun moves up and the setter flushes the grouse, allowing the gun two shots at the grouse as they fly off. It is good exercise, and was the method used when grouse shooting

started in the north of England and Scotland at the beginning of the nineteenth century.

In 1831 the Game Act was passed to protect the interests of all those preserving game for shooting, setting dates in the calendar between which it was permitted to shoot different species of bird. It also gave gamekeepers the power to carry weapons and to arrest poachers. This set in motion a chain of events that ensured that the goshawk had disappeared by 1889, the marsh harrier by 1898, the osprey by 1908, the honey buzzard by 1911 and the white-tailed eagle by 1916.

It was a prosperous time in England and rich sportsmen lusted after shooting increasing numbers of grouse, just as they shot driven pheasants flying out of a covert. This passion for driven grouse was aided midway through the nineteenth century by two factors: Queen Victoria's love affair with Scotland – and Balmoral in particular – and the expansion of the railways.

So how does driven grouse shooting work? The moor is divided up into a number of beats, and towards the end of each beat a line of six to eight butts – breast-high embrasures, behind which stand the guns – are positioned. A line of beaters, starting from the far end of the beat, walks slowly forward through the heather towards the butts. The red grouse, which tend to congregate in packs at the end of the breeding season, are flushed and fly very fast at shoulder height towards the guns. Good shots kill two birds in front, take their second gun from their loader, who is standing behind them to their right, and shoot

another two birds as they fly away. It is the kind of shooting that sorts the veterans out from the rookies.

Over time the grouse shooters demanded grouse in ever-increasing numbers. This led to overstocking, with more grouse left in the winter than the moor could sustain. Two diseases – *strongylosis* (caused by a worm in the gut) and looping-ill (a tick-borne disease of sheep) – found ready purchase in the weak birds and were to bedevil grouse-moor management for many years to come. Grouse disease was such a problem that the government stepped in and commissioned a two-volume monograph, *The Grouse in Health and Disease*, published in 1910. Dr Edward Wilson, one of the heroes of Scott's ill-fated attempt on the South Pole, was among the team that compiled the report.

They concentrated on *strongylosis* and succeeded in plotting out the disease's life cycle. Sometimes the disease can develop very rapidly, completing its cycle in fifteen days. The larvae hatch out in the droppings of the grouse and then climb up to the green shoots on the heather. The grouse eat the green shoots and larvae, then the larvae grow to their adult stage in the grouse's gut and reproduce, laying eggs. They pass out in the grouse's droppings for the cycle to be repeated. All red grouse carry *strongylosis* and late spring – April to May – is when the disease peaks. Allowing too large a stock of birds for the moor will lead to a periodic epidemic.

It is an interesting sidenote that in 1908 a protozoan parasite was discovered, *Cryptosporidium baileyi*, that

affects poultry. A hundred years later it would cause what is known as 'bulgy eye syndrome' in grouse, and had a devastating effect on driven grouse shooting when the moor was left overstocked at the end of the season.

Not long after the discovery of *Cryptosporidium baileyi*, driven grouse shooting and the number of grouse shot were reaching their apogee. I consulted *Record Bags and Shooting Records* by Hugh S. Gladstone for information on numbers of grouse shot, and discovered that on 12 August 1915 at the Little Abbeystead beat in the Forest of Bowland eight guns shot 2,929 red grouse in six drives, a record that stands to this day.

For several days the female hen harrier, number 22, tried to return to her birth place but was put off by the heather burning on the moorland below the ridge. Now, as she wheeled high above, she could see that the men had gone and all that was left was a mosaic pattern of blackened, burnt areas. Cautiously she drifted lower and then made several passes along the ridge until she was satisfied that it was safe. She dropped down and landed by the bilberry patch she remembered so well. The sky darkened as a ragged shower of rain swept across the moor, dousing those newly burnt heather patches, which were still smouldering.

Heather burning is part of the history of driven grouse shooting, a sport that started in about the middle of the nineteenth century. Beforehand moor owners had derived income from farmers wishing to graze their sheep, and the farmers burnt the heather to improve the grazing. Moor

owners realised that they could receive a better income from driven grouse shooting than from sheep grazing, so heather burning was banned. All went well for a few years until grouse numbers suddenly collapsed. What the moor owners hadn't realised was that red grouse eat heather shoots virtually throughout the year. Burning was quickly reinstated, with the aim of burning patches of heather on the moor when it was about ten years old. Burning at this age stimulates regeneration of growth from the roots to provide the green shoots that grouse feed on. If the heather is not burnt until it is fifteen or twenty years old there will be a very hot 'burn', caused by the long, woody heather stems – and there will be no regeneration.

The burning, which starts in October and continues until April, should produce a mosaic of burnt areas (an area burnt this year, last year and so on), the areas becoming greener and greener as the heather regenerates, with the edge of each burn ragged rather than straight. During the

first fortnight after they leave the nest, grouse chicks feed on insects and wander into the open, and ragged edges gives them a chance of escaping a hunting hen harrier while a straight edge leaves them totally exposed.

The sun was low over the horizon as the male hen harrier crossed the River Ribble and flew up into the Forest of Bowland. Smoke from heather burning was still drifting across the moors and fells, with men in lines tending the edge of the fire. He turned sharply to avoid them, as last year he'd lost three primary feathers to a shotgun blast. He flew straight to one of his favourite hunting areas, an open area of fell well covered with grass. He saw the occasional sheep grazing contentedly, but they took no notice of him. Eventually he found what he was looking for, a marshy area from which sprang a rill that eventually fed into the Ribble. He dropped down until he was about ten to fifteen feet off the ground and began a methodical search for prey into the wind. Flap, flap, flap, glide.

At the end of the marshy area he turned and drifted down with the wind, a V-shaped silhouette that hovered now and then to investigate a movement or a sound. By the rill were several stands of rushes. He heard a squeak and hovered over it, his ears hidden behind the owl-like mask of his face picking up the rustle of movement through the rushes. He extended his long yellow legs, closed his wings and dropped down. His eight black, needle-sharp talons unerringly grabbed the prey, a short-tailed field vole.

Short-tailed field voles thrive in any patch of land or field where there is rough, tussocky grass. Mainly nocturnal, their presence is indicated by holes in grass tussocks

where they nest, runs in the grass with holes where they pop up from time to time and remains of grass clippings that they have chewed. Both male and female field voles mark their territory and defend it with splashes of urine. It is estimated that there are 75 million field voles in the British Isles, and their breeding season starts in April or May, continuing through to September or October. A succession of litters is produced, with an average litter size of five, and young born at the start of the breeding season will themselves breed later in the same year. Their breeding success follows a cyclical pattern, a poor year followed by a better one. Then, either in the third or fourth year, there is a peak breeding season. This cyclical pattern is exploited by predators in the good years and has a depressing effect on their productivity during lean periods. Research has shown that peak vole years influence hen harrier numbers, with bigger clutch sizes and number of chicks fledged. Small mammals make up a modest but important part of the hen harrier's diet throughout the year, and two voles would meet the daily prey requirement of a harrier.

Volume 3 of *The Handbook of British Birds*, 2nd edition, which I bought as a young boy in 1944, gives a concise list of birds' prey items. The handbook states that the hen harrier:

> preys chiefly on birds and mammals taken by surprise on the ground. Mammals include young rabbits, leveret, mice, field- and water voles, rats, etc. Birds: frequently Meadow-Pipits, Sky-Larks, or young Lapwings; occasionally chicken or duck, Teal, Red Grouse, Partridge, Golden Plover, Snipe, Dunlin: also finches

> (Linnet, Chaffinch, Snow-Bunting), Song-Thrush, Blackbird, Ring-Ouzel and Stonechat. Snakes (Adder), lizards, slow-worm, frogs, etc., also taken, and eggs or young of ground-nesting birds (Meadow- Pipit, Dunlin, etc.): small fish also once recorded: also Coleoptera.

Eagles, Hawks and Falcons of the World by Leslie Brown and Dean Amadon, published in 1968, tackles the subject from a different perspective. The authors cite one analysis that gives the hen harrier diet as 25 per cent birds, 55 per cent mammals and 29 per cent snakes, frogs and insects. Another nesting analysis gives 31 per cent birds and 69 per cent mammals, with meadow mice predominating. Brown and Amadon state that the hen harrier can occasionally take mammals up to the size of a rabbit, and birds up to the size of a young bittern, but large birds such as ducks are usually wounded or moribund when taken. Food is usually taken on the ground, as the hen harrier can rarely capture birds in the air. Food requirements vary from 100 grams daily for a female (19 per cent of body weight) in cold weather to 42 grams (12 per cent of body weight) for a male in warm weather, the maximum daily intake by a female being 142 grams in cold weather.

Some of the more unlikely species that hen harriers have tackled include hedgehogs and adders, and apart from red grouse hen harriers have been known to take ptarmigan, black grouse, partridge and pheasant. Intra-guild predation has accounted for merlin and kestrel . . . and lastly, an unlucky short-eared owl.

CHAPTER FOUR

There was no sun and the moor was white with frost when, two days later, the male hen harrier drifted down to Mallowdale Pike ridge in search of a mate. He passed up and down a couple of times, and on the last flypast noticed a female hen harrier in the tall heather tearing at the tussocks of dead grasses below, throwing them in the air. Cautiously he planed down and landed close by. She was larger than him, and in the first glimmer of sunshine he admired her glossy chocolate-brown plumage, noticing the white wing tag with the number 22 *in black. He walked over to her. She looked at him and he could feel her sizing him up. Was he good enough? He'd show her.*

He rocketed up into the sky. The female watched as his blue-grey outline melded into the now blue sky above. At three hundred feet he turned and plunged into a vertical dive towards her. In freefall he corkscrewed and screamed to attract her attention. Faster and faster he fell. Then, just when it seemed that he would crash right into her, he pulled out and soared up in the air again. He was reckless in his efforts to impress her. Three times he repeated the death-defying plunge earthwards before finally landing by her side. He was panting, plumes of breath hanging in the frosty air. She was impressed and sidled over to crouch submissively by his side. He had found a mate.

'My routine when I'm checking for the harriers' arrival,' says Stephen Murphy, 'is to drive round to the north side of the central mass of the Forest of Bowland. Here it is split by a valley, and where the harriers nest is way back beyond the horizon. I'll park my car and watch. This is a muster point for hen harriers looking for a mate. In March

and April I have seen eight birds there, all "skydancing". It's like a big aerial dance floor, a real sight to behold. When they've found a mate, they'll fly up the valley and up out of sight to where they'll find a nest site.

'A week or so later I'll walk up onto the fell and find myself a spot in the heather where I can scan the distant hillside through my binoculars. At the moment I'm checking one of the traditional harrier nesting sites way across the valley. It's a favourite site – they've always produced young from there. We're in the last couple of weeks of March and any pairs of hen harriers should be arriving any day now.'

It is the female that makes the choice of where the nest will be, and shortly afterwards she can be spotted flying in carrying grasses or large sprigs of heather or bilberry in her beak, a sure sign that nest-building has started. Starting from scratch, the female chooses a bare area for the nest in tall heather but with easy access to it from one direction. Generally, all the nesting material – from quite bulky twigs to smaller sprays – is picked up within two hundred metres of the nest.

Flying out from her chosen site she pitched on an old heather burn and meticulously pulled up lengths of dead heather with her talons, then flew to the nest site. Here she started intertwining them. Off she went again, and each time she returned she methodically knitted the heather strands together, gradually building up the finished nest. She stamped around in it to perfect its saucer shape, then lined it with grasses and, as a final touch, added some

bilberry leaves. As she did so her white wing tag with the number 22 in black came into view. She noticed that she was losing feathers on either side of her breast-bone and that the bare areas were becoming suffused with blood – these were her brood patches.

Three other pairs of hen harriers had arrived on Mallowdale Pike, and were busy skydancing and searching for nest sites, making it a communal nest site.

Stephen Murphy lowers his binoculars. 'I remember helping my friend David Souter wing-tag that bird last year,' he said. 'This was the area from which she fledged. She's obviously decided it's where she's going to nest.

'I always keep well back from a potential nest site, at least 500 metres. With a good pair of binoculars you can be pretty sure of what's happening, and it all goes in my diary – weather, behaviour and so on.'

Their nest complete, the pair were ready for the next stage of their courtship, the magic of copulation. Seizing the moment, the cock bird flew in with a vole he had just caught. The female rose up from her nest and caught it as her mate passed it to her, and then they both landed. The female called to the cock bird, which approached her in a frenzy of excitement and mounted her. She moved her tail to one side, and he manoeuvred so that their engorged sexual organs came into contact. He flapped his wings to keep in position, there was a brief shudder and it was all over.

For the next week or so the pair of harriers copulated several times a day, before his mate became moody and disinterested and took to the nest she had built, these last acts of copulation having

stimulated a daily release of eggs from the ovary. It was now the middle of April. Nearby a hen stonechat had built her nest of moss and grasses at the bottom of a thick stand of heather that was just starting to sprout some green shoots; she was incubating five speckled brown eggs. Her mate was perched on the top of a branch of heather, asserting his right to the patch by constantly spreading his tail and flicking his wings.

On the fifth night the female hen harrier became restless, shifting uneasily in the nest. Finally, at four o'clock in the morning she stood up, legs well apart, arched her back and laid an egg. It was pale blue. She peered at it, touched it with her beak and then settled down over it. At first light the cock bird dropped into the nest and offered her a meadow pipit. She shuffled off the nest to eat it. Her talons grasped the pipit, her beak tore it apart

and she gulped it down, allowing her mate a moment to proudly inspect the newly laid egg.

Roughly every forty-eight hours she laid another egg, until she had a full clutch of six. A day or two after laying, the pale blue colour of the eggs changed to a chalky white. The inflamed bare areas on her breast were now hot to the touch and, with the arrival of the second egg, she started incubating. When she left the nest it was only to fly around for ten minutes or so before her hormones pulled her back to her treasured clutch of eggs. The cock bird was very attentive, bringing her food to the nest and roosting nearby at night. Incubation was a long process, and it would be another three and a half weeks before the first egg would hatch.

Suddenly, the female heard the cock bird calling. She took off and flew up to meet her mate, who was flying towards her. He was carrying a meadow pipit, and he slowed down to enable the female to turn and trail below him. He dropped his prey and the female flipped over in the air and caught it.

Stephen Murphy crouches in the heather, watching. 'After the food pass the female flew to a boulder to eat the prey item before returning to her nest, a definite confirmation of where the nest is. I don't want you to think that I'm working alone all the time. I'm fortunate to have the back-up of RSPB watchers who are employed on a summer season-only basis and I'm also able to call on local keen bird watchers whom I trust. We share any information we gather.

'In 1975, when there was only one pair of hen harriers nesting at Bowland, the two Bills – Bill Hesketh and Bill

Murphy – mounted a twenty-four hour watch on the nest site to ensure that the young were reared successfully. In fact, in 2011 it was the two Bills who found the nest that produced Bowland Beth.'

'It was 6 April,' says Bill Hesketh, 'a glorious morning on my watch as we settled down among the heather in the shadow of a peat bank. The whole of wild Bowland proper laid out in front of us – just heather, bilberry, rush and sky. At 10.45 am, quite a distance away, a female hen harrier came floating down from Mallowdale Pike. Her right wing had a tag on it, white except for a narrow pink band at the base – no tag could be seen on the other wing. Over the next fifteen minutes she began criss-crossing last year's nest location at a low height, at intervals setting herself down on the same spot.'

CHAPTER FIVE

Predator control of hen harriers by gamekeepers up until the Second World War was indiscriminate. The shotgun, the pole trap and poison were their weapons. It eased as gamekeepers joined up to fight in the war, and the hen harrier population recovered. The Protection of Birds Act 1954 gave full protection to the hen harrier and all other birds of prey – apart from the sparrowhawk. In 1962 the original Act was modified to include the sparrowhawk.

But the regime of heather burning revived the red grouse population and with it the popularity of driven grouse shooting. Persecution of all birds of prey, particularly the hen harrier, increased dramatically.

In 1981 the young at six hen harrier nests were wilfully destroyed, leading to a huge outcry. As a result the North West Water Authority and United Utilities joined forces to support an RSPB presence in Bowland, led by John Armitage.

'The evidence was just left there,' says Armitage. 'It was quite blatant. Nowadays it would have been retrieved, covered up. Running alongside this persecution there was egg collecting. There were about forty breeding pairs of hen harriers in Bowland, which was very convenient for egg collectors based in England. I started a dialogue with the four main estates to gently remind them that we were looking over their shoulders and would take action if we had the necessary evidence. I found that several of the keepers hated hen harriers – they couldn't even say the name and wouldn't speak openly about them at all.'

During the period between 1981 and 2005 John

Armitage correlated the successful nesting attempts on four of the Bowland grouse moors. It makes for very interesting reading: NWWA/United Utilities 153, Bleasdale 37, Abbeystead 34 and Clapham 15.

'I was disappointed that the RSPB didn't put its foot down and persuade United Utilities to give up grouse shooting on their land,' he says. 'As it stood then – and still does – it makes it easier for people to come in from elsewhere and clear hen harriers out.'

'The traditional way of dealing with hen harriers was to have a coordinated strike on all the roosts throughout the Pennine chain of grouse moors,' says Bill Hesketh. 'Mist nets would be set up beforehand, concealed on the roost. At dusk they would watch the harriers settling in for the night. At a prearranged time the nets would be pulled upright, a shot fired, and the keepers would rush in. Any harriers caught would have their necks promptly wrung.'

Another method used by keepers was to crouch over a nest containing young. The adults would dive bomb the intruder, getting closer and closer with each attack. It was crude, but simple and silent – just whack them with a stick as they passed by.

Terry Pickford has spent most of his life trying to safeguard the peregrine falcons and hen harriers on the Forest of Bowland. 'The late Duke of Westminster, when he bought the Abbeystead grouse moor, was determined to get rid of hen harriers. He employed young gamekeepers who could run faster than we could, and the persecution was blatant. First of all he eradicated all the harriers on

his moor and then they started disappearing from other grouse moors.'

Bill Hesketh describes a more sophisticated method of destroying hen harriers: 'On 22 May 1994, while walking on Whitendale Fell, I came across a long line of green twine well hidden in the vegetation. At the far end of the line I found a full-size replica of a male hen harrier made in plastic. I immediately realised that this was a lethal device designed to lure harriers to within gunshot range. I'd probably disturbed the keeper just as he was about to deploy it. The device was handed into the RSPB headquarters for investigation.'

All over the moor, another bird – red grouse – inextricably linked to the hen harrier, is laying claim to

their breeding territory. They had started late, as it had been an incredibly severe winter. Nevertheless, cock red grouse, a very striking bird indeed, are on show all over the moor, standing bolt upright, red wattles wobbling, uttering their familiar cry – *go-bak, go-bak* – to any intruding male. The female, with her duller plumage, makes a very rudimentary nest – just a scrape in the ground, although well hidden in the heather – in which seven or eight eggs are laid. They are handsome eggs, light brown in colour with dark chestnut splodges laid at two day intervals. Incubation starts when the last egg is laid and normally lasts about twenty-four days. During this period the hen only leaves the nest to feed, slinking away through the heather to be joined by the cock bird. He sticks close by her, watching over her as she feeds on the bright green heather shoots until she returns to the nest.

CHAPTER SIX

Except when he was hunting, the male hen harrier was always loitering over the nest site, keeping an eye out for intruders. One day, about halfway through the incubation period, he spotted a group of women – botanists, with the permission of the local landowner – moving up the fell towards the nest site. They were clad for the conditions: walking boots, windproof trousers and parkas. Their leader was wearing a knitted woollen hat with a pom-pom dangling from it. Now and then they stopped to cluster round a specimen that had attracted their attention and discuss it.

As they moved on uphill the male hen harrier decided that they were getting too close for comfort. He winged over and dived on the group, yikkering, skimming over them before pulling up sharply. They panicked, running back downhill. He dived on them again and again. On his final dive he pressed down harder, singling out the leader of the party. She grabbed her bobbled hat to save it from his talons. As he resumed his station high over Mallowdale Pike he watched the botanists scuttle away into the distance.

Stephen Murphy crouches at his usual viewing point. In the distance he sees a couple of walkers approaching. 'If I see anyone walking on the fell that might disturb the harriers,' he says, 'I cut them off and politely tell them to change direction – and why. Most walkers understand my concern.'

Insidious tactics carried out by grouse-moor owners to deter hen harriers from nesting on the moor include taking stock of grouse numbers as late as possible and heather burning continuing right up to the limit of 15 April. Stephen Murphy has often seen harriers wreathed in smoke from heather burning as they scour the moors for a safe nesting site.

It was now over thirty days since the female had started incubating. One evening as she turned the eggs she paused, cocked her head to one side and listened. A faint peeping was coming from within the egg. She settled down to brood again. On the following day more peeping, and a tiny crack appeared in the shell. The female softly called encouragement. The chick was using its 'egg tooth', a sharp protrusion on the upper side of its beak, to chip open the shell. Finally, in the early hours of a May morning, the shell split open and an exhausted chick, later to be known as Bowland Beth, flopped into the world.

Her tiny body was covered with thin pink down. Her eyes were closed and her head seemed enormous in relation to her body. She was cold, no longer swaddled in the comfort of the egg. Sound filtered in to her: the wind stirring the bilberry branches, the plaintive call of a golden plover. Her mother looked proudly down at her. For the next four or five weeks all her energy and love – her whole world – would be devoted to raising her brood. She picked up the two pieces of shell and flew off to dump them well away from the nest before returning to cover her newly hatched chick and resume incubation. For the next fortnight she would remain at the nest, brooding her chick, waiting for the other eggs to hatch and relying on the cock bird to bring in food regularly.

Sadly, two of the eggs failed to hatch. When the remaining four chicks were over a week old, it was felt safe for the RSPB technician to place a CCTV camera on the nest site. These cameras are non-intrusive, film in high-definition and produce an excellent record of behaviour at the nest.

'The four young are growing day by day,' reports Bill

Hesketh. 'The eldest are well feathered and practising flapping their wings and investigating their reflections in the camera lens.'

CHAPTER SEVEN

Out on the moor the first red grouse chicks are leaving the nest a few hours after hatching. They are buff coloured, with a disruptive pattern of darker streaks. For the first two weeks of their life they feed on insects, their favourites being daddy-longlegs, small flies and click beetles. This food, and particularly the protein in it, enables the chicks to grow quickly to a size where they can fly. If the nest area is very dry the hen will lead them to wet areas with mosses and rushes where there will be a plentiful supply of insects.

During this fortnight before the chicks can fly, the cock bird is always on the lookout for intruders. If he senses danger he will put on an injury-feigning display to distract attention from the young, flapping along the tops of the heather as though he's got a broken wing.

It had rained all day. The female was mantling over the four chicks, brooding them, keeping them dry. She was soaked, her feathers in disarray, those on her head standing up in spikes. When the sun came out she stood up and shook herself, throwing a spray of sparkling spangles across the nest onto her four chicks huddled together. At fourteen days old they were now clad in thicker white down and had their heads up, eyes wide open. She cocked her head. She'd heard her mate calling. She sped up to fly just beneath him. As he dropped the prey item she flipped over on her back and caught it in an outstretched talon. The chicks crowded round as she returned to the nest. She tore open the meadow pipit and called to make them gape so that she could thrust tiny parcels of meat into their open beaks.

Bowland Beth was the biggest and most adventurous of the

chicks, and she pushed forward greedily to get more than her fair share. The female ignored her, feeding each of her chicks in turn until all their crops were bulging. While the chicks flopped down asleep she busied herself clearing up debris, swallowing leg bones that the chicks would not be able to cope with for another fortnight or so.

CHAPTER EIGHT

Lowering dark clouds shrouded the heather moorland. The rising sun broke through, flashing silver on an area dominated by grasses and rushes. The male hen harrier was hunting for the hungry chicks but so far had only brought in small items of prey – meadow pipits, skylarks and stonechats. Now that the chicks were over three weeks old they needed something bigger, such as plump grouse chicks. Flap, flap, flap, glide. It was a relentless search pattern, up into the wind, then back downwind.

Above and behind, and dwarfed by the hen harrier, was a cock merlin, the smallest British bird of prey. Its upper plumage from head to tail was blue-grey, its underparts pale brown with dark longitudinal streaks that thickened towards the belly and flanks. It had a curious relationship with the hen harrier. It also had a nest nearby and was very protective of it, noisily repelling any intruders. In exchange for this early-warning system, the hen harrier tolerated its presence and allowed it to hunt alongside, picking up any flushed prey that eluded him.

The male hen harrier suddenly veered away from the merlin as he had seen something in the heather below. He spread his tail, flapped his wings, then stopped dead in the air before plunging down, legs outstretched, to drop on the unsuspecting prey, a grouse chick. The remainder of the brood escaped, fluttering away into the heather.

At the nest the female was wholly dedicated to caring for her four chicks, and she looked around anxiously for the return of her mate with food. The chicks were developing well and they were beginning to be aware of the world outside the nest. The largest chick, Bowland Beth, was standing upright and investigating a beetle scurrying across the nest, while the others rested on their

hunkers. Suddenly the female heard the cock bird calling. She looked up as he dropped the dead grouse chick into the nest.

The chicks ran in to gather round the female as she grabbed it. She plucked the bird and then butchered it neatly, giving each of them bloody gobbets of flesh until they were sated. Three of the chicks collapsed in a heap, the fourth headed for the edge of the nest and, quite out of instinct, defecated deep into the heather.

At the nest site there had been a transformation over the previous week. The four chicks – two males and two females – were now four weeks old and starting to look like young harriers, their emerging back and breast feathers replacing the down. They had become increasingly active, flapping their wings and scrambling into the tunnels that they had made in the tall, green-fringed heather that surrounded the nest.

Suddenly they all squatted down as a shadow loomed over the nest, followed by a hand that took one of the chicks away. They were being ringed by an official ringer. One by one, a thin metal ring with a six-figure number engraved on it and an address to which it should be sent if found was fitted to one of their legs. The ringer also measured the length of the fifth or sixth primaries from the carpal joint to the tip of the feather. It was a frightening experience for them, their first encounter with man. As each chick was placed back in the nest it scurried off to safety in the tunnels among the thick heather. When the female returned with food half an hour later they rushed out to surround her, the moment of danger forgotten.

'It's 20 June,' says Bill Hesketh, 'and the four young are growing day by day. The eldest are well feathered, and

practising flapping their wings and investigating their reflections in the camera lens.'

'I happened to be watching when the official BTO ringer came by to ring the harrier chicks,' adds Stephen Murphy. 'I asked him to let me know the length of the primaries so that I'd know when to fit the satellite tags.'

The origin of ringing birds can be traced back to the Reverend Gilbert White. He tied a cotton ring around the leg of a swallow to try to find out whether it would return to the same nest the following year. The metal rings in use today were pioneered in Britain in two schemes by Harry Witherby and Landsborough Thomson that were amalgamated into the British Trust for Ornithology in 1937 and continue to the present day. An additional coloured ring with letters enables individual birds to be identified. Wing-tagging – a white or coloured thin plastic tag with numbers or letters on it – was started in the mid-1990s. A few days before fledging the tag is clipped onto the soft patagial strip on the leading edge of the bird's wing. The procedure is not without its detractors; some experts think that the tags attract other predators' attention and they cite well-documented evidence of peregrines taking out wing-tagged hen harriers in Spain and England.

'When I started at Bowland in 2002,' says Stephen Murphy, 'we put radio tags on the hen harrier chicks. They were held in place by a lightweight harness that fitted over the wings. These tags gave us interesting information. I remember being phoned by Kjeld Jenson, who told me

that he'd seen one of our cock birds trailing a Danish hen harrier on the Isle of Als off the coast of Denmark. Radio tags were OK, but it was evident that there was still much to learn.

'The radio tags were useful for tracking birds from Bowland as they made their way up onto the grouse moors. One day I was tracking a couple of harriers with my Yagi aerial when a jeep pulled up beside me. A couple of police officers from RAF Menwith Hill, a top-secret listening station, got out and asked me what I was doing. I said I was tracking hen harriers. They said either you're a liar or that's the best story we've ever heard. Luckily, one of the harriers we were tracking suddenly came within range so they could hear the signal. I obviously won them over because, from then on, they regularly sent me sightings of any of our birds they picked up.

'In 2007 radio tags, which only lasted for eighteen months, were phased out and we turned to satellite tags. I'll be fitting one on one of the chicks in three or four days' time.'

CHAPTER NINE

Three days later the female, number 22, stood watching her brood, full of admiration for them. Suddenly she flapped off as she heard the cock merlin's alarm call. All the chicks took cover in the heather.

The cock merlin was stooping, time and time again, at something moving through the tall heather. At the end of each stoop he looked back over his shoulder and screamed at the unseen intruder. The female hen harrier appeared on the scene and joined in. A particularly fierce stoop from her flushed the intruder out into the open. It was a fox. Working in tandem the female hen harrier and the feisty cock merlin hit him hard. Ears laid back, the fox made a hasty retreat away from Mallowdale Pike ridge.

Stephen Murphy has an interesting theory about why foxes are drawn in to hen harrier nests. 'You go to some harrier nests,' he says, 'and they are antiseptically clean, not even an entrail, a foot. Nothing. These nests belong to birds where the females have to do a lot of the hunting themselves and they are probably getting very little help from the males. Or they are bad hunters, bad parents. They can't bring in the amount of food the young really need. On the other hand the nests of the really good hunters are rank. Here you will find birds lying around the nest in different stages of decomposition; half-eaten grouse, bits of meadow pipits, small mammals, that sort of thing. When the well-nourished and developed chicks are twenty-five days old their appetite wanes, but some parents keep on bringing in food. No wonder foxes are attracted to these nests.

'In the intensely keepered moors of Bowland fox density is low, and to the best of my knowledge no hen

harrier nests were lost to fox predation between 2002 and 2015. This would also apply to most grouse in the Pennine chain. After they gave up predator control at Langholm, to the north-east of Carlisle, they installed a CCTV camera on the nest when a fox came in to predate the chicks. The older chicks put on a good show, raising their wings up in the air, hissing, making themselves look as big and ferocious as they could. The fox scarpered off, discretion being the better part of valour.'

At five weeks old, in the third week of June, the chicks were now full grown. They were all standing, some preening away the last flecks of down. One was desperately trying to swallow a grouse chick's leg. She kept gulping but it wouldn't go down. Bowland Beth ignored her sibling's antics. She was energetically wing-flapping, almost taking off. She looked ready to fledge. There was something extra special about her. She glowed like a newly minted coin and had the aura of an Olympic athlete.

'Kek-kek-kek-kek.' The female hen harrier's yikkering alarm call from above froze the chick's activities, and they both crouched in anticipation – there was to be another encounter with man. As before, there was the crunch of footsteps in the heather before a shadow loomed over the crouching chicks.

'As we advanced on the nest,' says Stephen Murphy, 'one of the chicks, the eldest male, sprang up from it and vanished. I picked up the next largest and carried her away from the nest so that I could fit her with a satellite tag.' Stephen sat down on the heather to carry out the

procedure, a helper beside him. They both admired this feisty bird whose beak was wide open in protest, her talons digging into Stephen's hand.

'My helper on that day was Jude Lane, the RSPB leader of the hen harrier project in the Forest of Bowland, and it was Jude and her volunteers who named her Bowland Betty after Bet Lynch, the character in the ITV soap opera *Coronation Street.'* (Despite this, Stephen always called her Beth, and I have followed his lead.) 'Bowland Beth was one of those birds that you come across now and then that are absolutely perfect – her plumage, a rich, glossy chocolate brown, the most wonderful eyes, yellow irises, and vivid chrome-yellow legs with formidable black talons.

She's what I call a "super" hen harrier.

'She obviously took after her father, who was now relentlessly dive bombing us, making us duck each time. I saw that he was holding a vole, and he must have thought, "I'm not going to jettison this. I've spent hours going up and down the moor to get it and nobody else is going to have it."'

Steve and Jude eased the harness over either side of Bowland Beth's wings to enable the satellite tag to be held firmly in position on her back. 'I set the satellite tag on a 10:48 pattern,' he recalls. 'That meant it transmitted for ten hours and spent forty-eight hours recharging in daylight. So there was a shadow when you didn't know exactly where she was but hoped that she was in the place you last had a fix on her.'

Bowland Beth hissed and scratched as she was put back in the nest. For a moment she lay there hissing, eyes glaring, beak wide open, her talons – needle-sharp scimitars – grabbing in the air. Gradually she calmed down, righted herself and gave her feathers a good shake. She noticed that one of her brothers in the terror of the moment had fled the nest. She decided she would stay, along with her sister and remaining brother, and wait for food to be brought in.

Stephen and his helper retreated back up the hill to resume their watch from a distance. Within minutes of settling down to watch they saw the absconder return to join his siblings.

CHAPTER TEN

Heavy rain fell for several days, then it cleared. Bowland Beth had recovered from being satellite-tagged and was wing-flapping furiously with her brother and sister. Facing into the wind, it looked as though she could fly at any moment. Later on, during an even more strenuous flapping session, a sudden gust of wind caught her and she was airborne.

After an initial moment of panic, flying became an automatic reflex. Her wings rose up and then drove down and backwards, pushing her forward through the air. She found that by flapping harder with one wing or the other she could turn to the left or right. Even flexing one primary made for a change of direction. Climbing was more difficult. She realised that she had to flap harder with her wings and, to avoid getting lost, had to use her tail to steer so that she was climbing upwards in circles.

She was at about two hundred feet now. Below was the Mallowdale Pike ridge, and flying up to her was her mother, with the number 22 *in black on her white wing tags. She led Beth back to the nest, showing her how to lose height gradually by partially closing her wings, steering with them so that she descended in a zig-zag and then fanning her tail to make a controlled landing on the ground.*

The three other young harriers were standing in the nest, hungry and waiting for food. Out in the distance the male hen harrier appeared carrying prey, calling. Beth flew up to join him with her siblings not far behind. When they were about fifty feet beneath him he dropped the prey item. Like a cricketer running in to catch a 'skier', she deftly manoeuvred into position, turned on her back and caught the prey with one outstretched taloned foot. She planed down to the ground, where she mantled over her catch

– a vole. Her siblings gathered round, keening for a share of her prey.

Over the next fortnight Beth's flying ability improved dramatically. On 23 July – a warm, sultry day – she latched onto a thermal and felt herself soaring up with very little effort. In no time at all she could see the sea to the west and, as she wheeled round to the north, a much bigger expanse of heather moorland came into view. Perhaps with her phenomenal eyesight, eight times sharper than our own, she saw something that triggered a response from her ancestral memory.

Beth immediately forgot the pleasure of soaring and set a course for the moorland known as Nidderdale, which was only thirty-five miles away. It was the start of a great adventure, a journey that would be followed with joy and alarm by a privileged few through the signals transmitted by the satellite tag on her back.

Just under an hour later Beth arrived over Grimwith Reservoir, the large expanse of water she had seen when soaring over Bowland. She crossed the dam, maintained her height and circled round, taking in the wide expanse of grouse moor beyond. She saw rows of grouse butts laid out as at Bowland, although she did not yet know what their purpose was. There were three streams on either side of her flight line that were emptying water off the moor into the reservoir, then there were more grouse butts. She steered clear of the butts and followed the middle stream, Gate Up Gill, to its source on Combes Hill.

After two circuits she landed at the top of the escarpment on Wolfrey Crag, a thirty-foot-high limestone cliff deeply fissured by run-off water. This gave a perfect view down the valley to the reservoir below. She flapped up to a thick stand of bilberry that

made a perfect resting and lookout perch, and there was thick heather all around in which she could shelter. Now routine took over. She started to preen, paying particular attention to the primary feathers in her wings and her tail feathers, drawing them through her beak to realign the barbs and anointing them with oil from the gland at the base of her tail. She fluffed up her plumage and gave it a good shake.

I ask Stephen what he thought had prompted Beth's sudden departure. 'Most of the birds we've tracked seem to want to get away from where they were fledged,' he says. 'Nidderdale is proper heather moorland, so naturally it would be attractive to harriers. It's a lush place holding a plentiful supply of meadow pipits. There'll be baby meadow pipits – little bundles of fluff, yellow round their beaks – hopping across the heather. Two easy prey items for a newly fledged juvenile hen harrier.'

CHAPTER ELEVEN

Then, like Robinson Crusoe, Beth took stock of her situation. On the debit side she was alone. She had seen no other harriers, although she did not know that her kind had always been persecuted. Her only friends were those who found pleasure in seeing birds of prey alive, admiring them as marvels of evolution and fighting to protect them. On the plus side she had just got the keys to the world and was as fat as butter, as she'd always been more alert than her siblings when her parents delivered food to the nest. Her mother had dropped in grouse poults that were still alive, so Beth had to catch them and deliver the coup de grâce. She had no need to go hunting today, but later she must find a place to roost.

The sun had slipped down behind the heather-covered hills that nestled round Grimwith Reservoir when Beth took off from Great Wolfrey Crag and made a wide circuit of Nidderdale Moor. A hazy harvest moon broke the skyline as she followed Ashford Beck, which wended its way downhill towards Pateley Bridge village. Turning back, she saw three harriers beneath her dropping down onto a boggy area at Flowery Wham. She watched as they circled the area, checking it out, before they dropped down.

Beth cut her speed and made a low pass over the saucer-shaped depression. As the light waned she could just make out the shape of the three harriers roosting in the bed of rushes, Molinia *grass and sphagnum moss. She flew past again, lower this time, and as she banked she flapped her wings to cut her speed, spreading her tail so that she was almost stationary when she dropped into the roost site. She looked around. With a warm billet and others of her kind as company, she tucked her head under her wing and slept.*

Hen harriers are sociable birds. If circumstances permit they nest communally and gather in the evening to roost. Harrier roost sites are generally situated in a wet, boggy area, giving them advance warning of impending danger. Roosting communally allows experienced members of the roost to share information on the best foraging grounds with inexperienced juveniles. The first to leave the roost know where a flock of winter thrushes was seen dropping in at last light. The juveniles follow their lead and head in the same direction.

When the three harriers left the roost at sunrise the next morning Beth tried to follow them. She was hungry. The adult female was having none of it, however, and with a quick dart at her sent her packing. Beth slumped down into the heather to collect her thoughts. Casting her mind back she remembered how her mother foraged for her and her siblings back at Bowland. She could manage. She took off and decided to follow the stream she'd seen the previous evening – Ashford Beck. Just as she had seen her mother hunt, she slowly flew along the heather edge into the wind. Flap, flap, flap, glide. Beth floated ten feet off the ground.

Her owl-like face with its yellow eyes and ears looked and listened for any movement that might indicate prey. How on earth was she going to forage successfully, with just heather and more heather whizzing underneath her? She changed direction slightly so that she was hunting over the clumps of Molinia *grass on the edge of the beck. But what was that? She hung in the air, tail spread, wings beating silently. Was there a movement behind that clump? She plunged down, legs outstretched, talons spread. It was a brown hare! For a moment their eyes locked and Beth saw herself reflected in the hare's pupils. Then it pulled itself together, lashed out with its hind legs and lolloped off.*

Thoroughly shaken, Beth left the beck and decided to resume foraging along the edge of the heather burn. Methodically, she flew low into the wind, searching the heather, then doubled back downwind, her wings held high, a floating V-shape. She had a split-second glimpse of a meadow pipit crouching. Braking hard with her tail and wings, she pirouetted in the air before crashing down into the heather. Her eight talons spread out, then clenched convulsively in a death grip. Mantling over the meadow pipit

she made short work of stripping away its plumage. Her beak slashed into its breast with the dexterity of a surgeon. Legs, bare of feathers, were gulped down. After ten minutes all that was left was the pipit's head and some bloody feathers. Her first kill on her own. Her mother would have been proud of her.

Throughout the rest of the day Beth accounted for two more pipits and a vole. She drank thirstily at the beck and ducked herself in the water before returning to yesterday's lookout perch on the bilberry patch at the top of Great Wolfrey Crag to preen and rest.

CHAPTER TWELVE

Beth was faring well. She was way ahead of her siblings, who were still hanging around their nest site and dependent on their parents for food. During the day young hen harrier fledglings stay scattered about in the heather. Their parents feed them at ever-increasing intervals, almost starving them so that they have to go out and fend for themselves. Eventually, after about a month, they will leave the nest area to become self-sustaining foragers.

Their young lives are full of hazards: they need to be able to contend with prolonged periods of bad weather; they run the risk of being shot; in the excitement of chasing prey they may collide with vehicles or overhead wires; ground predators sometimes catch them unawares at their roost sites; and intra-guild predation by buzzards and goshawks may also be responsible for culling weak and unwary juveniles. The life expectancy of a juvenile hen harrier is not great – only one of the four young fledged from the nest at Mallowdale Pike would be lucky to be alive a year later.

'I would not think any bird of prey is safe in Nidderdale,' says Guy Shorrock, the RSPB's senior investigation officer. 'The whole of north Yorkshire has had the worst persecution record of any county in England or Wales over the past twenty-five years. Most hen harrier breeding attempts have failed, and Natural England assesses this as due to persecution. It's not a healthy place for a hen harrier to be.'

Beth left the roost at Flowery Wham just before dawn. Gripped by hunger, she decided to forage along Ashford Beck where she had killed successfully the previous day. She flew low in the

gloom along the edge between heather and grass, surprising a young rabbit. Dropping on the creature, she grabbed it across the loins with one taloned foot and grasped its head with the other. It squealed. Beth squeezed its body convulsively, the rabbit went limp and she pulled it under the cover of the heather. She plucked the rabbit's fur, throwing it in the air, and it blew away over the heather, eventually falling into the beck. At last the steaming carcass was exposed. She broke into it and took her fill.

Later, her full crop protruding like a pouter pigeon, she flew on to her lookout perch at the top of Great Wolfrey Crag to digest her meal and drifted off to sleep.

She was awoken by a fusillade of gunfire. She left the safety of Great Wolfrey Crag and headed east, found a thermal and soared up high to look down on Heathfield Moor. Below, a line of about fifteen men were walking across the moor. Those at either end of the line carried white flags, which they waved. The head keeper in the centre was urging the beaters not to straggle, to keep in line. As they advanced, packs of red grouse flew up and skimmed forward over the moor towards the line of eight butts strung across their line of flight. Men in the butts clad in full shooting garb – flat hats, jackets and plus fours, and brightly coloured stockings and garters – raised their guns and shot at the approaching speeding grouse. They handed their discharged gun to their loader, took their second gun and fired again. Now and then grouse crumpled up in the air with a bomb burst of feathers before dropping into the heather.

It was the Glorious Twelfth – 12 *August – the start of the grouse-shooting season.*

A horn sounded when the line of beaters came within sight of the guns. Shooting stopped. Dogs scurried out, tails wagging, to retrieve

the fallen birds, which were laid out in lines behind the butts.

Ten minutes later, Beth watched the guns being driven to the second line of butts, ready for the next drive. The beaters lined up again across the moor, the horn was blown and the drive began. Beth maintained her height and followed the proceedings covertly.

As she watched, the three hen harriers she had roosted with the previous evening appeared flying low over the moor. Grouse that were being driven forward by the beaters either veered off or dropped into the heather.

Her innate fear of man, even though she was soaring at over a thousand feet, made her pull down with her left wing, feather her tail and cut free from the thermal. Dropping like a stone, she was soon safe at Great Wolfrey Crag.

Back on the grouse moor fists were raised and shaken at the intruding hen harrier. A figure, not so garishly dressed and trailed by retriever dogs – he was obviously the owner of the moor – hurried over to the man who had blown the horn and was in charge of the beaters. He was the head keeper. There was a terse conversation before the owner returned to his guests.

In just an hour on the grouse moor Beth had quickly learnt one of the facts of life: that man loved shooting red grouse and hated hen harriers because they spoiled his sport and ate his grouse.

'Packs of grouse usually fly when they see the approach of any broad-winged predator, such as an eagle or harrier, unless it is high in the sky,' writes Donald Watson in his book on the hen harrier. When individual guns have paid handsomely to shoot driven grouse, the disruption of a drive by hen harriers is a serious matter, a great loss of face

to the moor owner and probably more irritating to him than the number of grouse chicks taken by harriers to feed their young during their breeding season.

Hen harriers do not commonly kill red grouse. Adam Watson and David Jenkins described in Donald Watson's book 'how a cock red grouse, when stooped at by a male hen harrier, stood up and struck towards the attacker with its bill, repeating this behaviour when the harrier stooped a second time. The harrier then flew on. On another occasion a hen red grouse flattened herself into the long heather beneath a stooping harrier. She then jumped up with wings spread and avoided the harrier's strike by a "quick bouncing movement" back to the ground.' It is probably only female hen harriers that are strong enough to take adult red grouse, and then only in winter when they are undernourished or sickly. A sustainable daily diet for a hen harrier would consist of either four meadow pipits or two short-tailed field voles. A winged or weak red grouse would keep a hen harrier going for two days.

All other birds of prey, including golden eagles and peregrine falcons, have a much wider choice of habitats and range of prey. The majestic swoop of the golden eagle and the 200 mph stoop of the peregrine falcon stir the admiration of even the most hard-hearted gamekeeper. The silhouette of a peregrine in the sky makes a grouse squat in the heather, while the presence of a golden eagle or hen harrier makes them 'flush' from the moor. The low-level flap, flap, flap, glide of the hen harrier seems underhand and dastardly, so is it any wonder that most keepers have no remorse about destroying them?

CHAPTER THIRTEEN

The shooting, six drives in all, went on for the rest of the day. Beth remained safe in her refuge on Great Wolfrey Crag, but she would have to be even more careful from now on. Eventually, in late afternoon, the noise of gunfire ceased. The sound of vehicles starting up and moving off into the distance faded away, and the 'kronk-k' of a raven echoed up from the area covered by the last drive. It must surely now be safe for her to have a look.

She took off, climbed strongly and headed for Heathfield Moor. Wheeling over the moor, she looked down. There was the raven picking at something on the ground – a dead grouse that had not been retrieved. Carefully she lost height, using the energy from her descent to skim across the moor. The pupils in her yellow eyes widened and suddenly she dropped down, making a grab with her long-taloned legs at the movement she had spotted. It was a red grouse that had been winged during one of the drives. She administered the coup de grâce just as her mother had taught her, then straddled the bird so that she could rip the warm flesh off its breast. Feathers flew everywhere, floating away over the heather, and in less than fifteen minutes Beth had stripped the breast-bone clean. She wiped her beak on the ground and flapped heavily across the heather, gradually climbing to a safe height before turning in a wide arc to look down on the moor. She noticed the raven scavenging leftovers from her kill.

For the next forty-eight hours there was a lull in the grouse shooting and Beth was able to relax again. She hunted over her favourite Ashford Beck, foraging for voles and pipits, rested on her bilberry perch on Great Wolfrey Crag and returned both evenings to roost at Flowery Wham with the three other harriers.

As she flew in to the roost on the third evening, however, she

could sense that something was wrong. The three harriers were wheeling above and the adult was venting her anger with a yikkering alarm call. Beth dropped down for a closer look, flying in circles around her roosting place. The whole roost site had been destroyed, ripped apart by vehicles that had deliberately driven backwards and forwards over the sphagnum bog and rushes.

She remembered another possible roost site she had seen three days earlier at Hard Gate Moss. In the gathering gloom she flew at low level onto the adjoining moor. To her left and ahead of her she could pick out the site she remembered. She climbed up to fifty feet as she made a slow pass over the site. It was intact, just as she had recalled it. Cautiously she made a couple of low passes over the area. Perfect. She braked with her wings and tail, then settled into her chosen roost of sphagnum moss and rushes. As she settled in to her warm billet she heard the other three harriers fly in to join her, although they soon went their separate ways and she never saw them again.

Destruction of roost sites does happen. In 2014 Sky and Hope, two juvenile hen harriers, disappeared under suspicious circumstances. A friend, part of a team looking for the birds, told me that an area in which they were searching for the birds' bodies was a roost site. It had been destroyed by Argocats – amphibious all-terrain vehicles – going backwards and forwards, destroying the rushes and heather.

Over the next few weeks Beth settled into a routine. Every morning at dawn she would leave the Hard Gate Moss roost and go foraging. Occasionally, her routine was interrupted by driven

grouse shooting and she kept a low profile, perched on the clump of bilberry at the top of Great Wolfrey Crag.

During the remainder of August and through September the bulk of the prey she caught consisted of meadow pipits and short-tailed field voles. Sometimes she would spend all day catching a single vole, but she once caught three voles in twenty minutes. Her preferred hunting area was still Ashford Beck, and she caught a water vole and a snipe there one day. If she was lucky, after a day's driven grouse shooting she might occasionally pick up a winged grouse, and this would keep her going for the next couple of days.

The grouse moors on Nidderdale would have anything from five to fifteen days' driven grouse shooting in a season, although in a good year it might go up to twenty. The idea is to shoot until a sustainable stock of grouse is left on 10 December – the end of the shooting season – to overwinter on the moor.

CHAPTER FOURTEEN

At the end of September two events occurred that disrupted Beth's routine. First, storm clouds gathered and it rained relentlessly, day after day. She couldn't hunt and had to seek cover in the thick heather on Great Wolfrey Crag. Second, the meadow pipits vanished.

The British population of meadow pipits leaves high ground in September and migrates south to France, Portugal, Spain and Algeria. A few remain. At the same time, a wave of migrant birds from Iceland, Greenland and the Continent arrive in Britain and spend the winter here. It is an interesting point that in spring those pipits that elected to winter in Britain are thin and not at all good-looking, their plumage having been abraded by the heather, while those that wintered abroad return fat and fit, with their plumage sparkling in pristine condition.

Maybe it was the diminishing number of meadow pipits that prompted Beth to fly north-east to the North York Moors. On 18 and 19 October she roosted at Bransdale Moor, then the following day she returned to Nidderdale.

Now that one of her main food sources had disappeared, on her return Beth foraged over wider and wider areas not only on the grouse moors but on lowland areas as well. One crisp, frosty autumn day in October she left her roost and sailed down from the heather-covered moor to inspect the meadows around Grimwith Reservoir. She flew down the River Dibb, grazing sheep scattering at her approach. With the wind behind her and flying low under the cover of a dry stone wall, she was able to flip over the wall from

time to time in the hope of surprising prey on the other side.

Three times she tried this method of foraging . . . nothing. On the fourth covert flip her patience paid off. In front of her, running away, were about eight birds smaller than red grouse. They kept on running, unwilling to fly. They were red-legged partridges, bred and released for shooting, and Beth easily overhauled them. She stuck out a leg and grabbed one. There was a flurry of feathers as Beth dragged it into the corner of the field under the cover of the wall. Here she despatched it, plucked it and took her fill of the rich meat on the bird's breast.

'The tracks on the satellite maps and the reports from Mick Carroll in the field all showed that Beth was flourishing,' says Stephen Murphy. 'She had survived both the start of the driven grouse shooting season and having her first roost site destroyed.'

All night long the first of the autumn gales lashed the moor in Nidderdale where Beth was roosting at Hard Gate Moss. The howling of the wind woke her and she cocked her head to watch the ragged clouds racing across the moon. Blown with them were flocks of birds, an almost unending stream. She snuggled down lower in her roost as the wind sliced through the rushes around her.

In the morning the wind had abated and Beth was able to float down across the reservoir to forage along the River Dibb. Here were hawthorn hedges laden with bright red berries, and feasting on them were hundreds of chattering fieldfares and redwings, migrants fleeing from the harsh Scandinavian winter. She dropped down to ground level close in to the hedge, the winter thrushes

much too preoccupied – cramming their crops with the luscious fruit – to notice her stealthy approach.

At the last moment she spread her tail, braked and darted up among them. She shot out a taloned foot, grabbed a redwing with a hawthorn berry still in its bill and carried it off to the cover of a riverside spinney. She started plucking it but was immediately disappointed – it was thin as a rake, with hardly any flesh on it. The long flight from Scandinavia to England had taken its toll. At this rate she would need two more kills to keep her hunger pangs at bay.

She carried the redwing over the wall and continued plucking it under the cover of a clump of rushes. Once she had eaten the meagre fare she started hunting again, employing the same tactics as before, flying along the wall and flipping over. To her amazement this time she had competition. Two male hen harriers, dazzling in their silver-grey plumage, were also harassing the fieldfares and redwings.

Over the next two days, a female adult hen harrier joined Beth and the two cock birds as they followed the flock of foraging fieldfares and redwings from the reservoir down through Appletree Wick pasture. Eventually Beth and the female adult hen harrier returned to Nidderdale.

'Historically, ornithologists stated that all the hen harriers in winter migrated to the coast,' says Stephen Murphy. 'Those female hen harriers that are spotted on the coast are almost certainly migrants from Scandinavia, as are most cock birds seen there as well.

'Our female hen harriers definitely stay on the

moorland all winter and most of them stay above two hundred metres. The males do have a migratory trait and will travel as far as Spain or France in winter. We know all this through satellite tracking. For instance, McPedro, a first-year male, went all the way down to Spain before returning to Scotland the following spring. The following year he went off to Brittany in France. No sooner had he got there than he disappeared. His last satellite signal came from the middle of an orchard.'

CHAPTER FIFTEEN

A chill wind was blowing across the moor as Beth left the bilberry perch on Great Wolfrey Crag and made her way east to her familiar roosting place at Hard Gate Moss. Behind her a blood-red sun slowly sank beneath the dark outline of Grassington Moor. She made two check sweeps of the marshy, rush-covered area before descending to her usual roosting place. Shortly afterwards the adult female hen harrier she had met out foraging dropped in and settled down for the night.

Later Beth was woken by a distant haunting chorus of calls heading towards the roost. There was a full moon and, looking upwards, she could see the silhouettes of skein upon skein of geese flying south. They were white-fronted geese, winter migrants on their way to the wildfowl and wetlands reserve at Slimbridge on the River Severn estuary.

At dawn she woke to find the moor gripped hard by an overnight frost. As yet there was no sun, everywhere still blue and cold and silent, except for the faint sound of a dog barking down at Grimwith House. Beth's facial disc, back coverts, primaries and tail were rimed with hoarfrost, so she stood up and shook herself, regurgitating a pellet at the same time. She surveyed the scene; the sun was just about to rise from behind the Wig Stones on her left and the other harrier had already left the roost. Beth shook herself again and took off for Ashford Beck.

She planed down in the cold, still air directly into the glare of the rising sun and landed on the edge of the beck. When she went to drink at her usual pool of water her beak hit solid ice. She moved forward carefully, slipping once on a thin skim of ice, to a spot where the beck was running. There she drank her fill before flying off to her daytime roost on Great Wolfrey Crag. She made

wide circles in the air, gradually building up height, till she looked down on the frost-gripped moor. She checked her wheeling as she noticed an advancing line of smoke and fire. Men were beating the flames, just keeping them under control.

Heather burning can begin at any time during the grouse-shooting season. A good sunny and frosty day is an excellent time to start, with most of the burning being carried out after Christmas when the keepers have finished with the pheasant- and partridge shooting season. It is a vital practice if grouse stocks are to be kept stable, as is the tackling of the two diseases that attack grouse: *strongylosis* and looping-ill. If not treated, both can cause crashes in the grouse population, which can occur every four or five years.

The deadly part of the life cycle of the strongyle worm begins when the adult grouse, pecking at a heather shoot, inadvertently ingests one of the worms, which then burrows into the gut, causing bleeding, a reduction in digestive efficiency and a consequent loss of condition. Nowadays there are effective counter-measures, including a network of gritting stations set up throughout the moor. Grouse need grit to aid their digestion, and starting in October, high-strength, medicated grit counteracts the strongyle worm, proving very effective. The medicated grit is removed by 15 April, four months before shooting starts on 12 August.

Looping-ill is a virus carried by ticks that is passed from a host mammal, generally sheep, to grouse. The virus is extremely virulent, with 80 per cent of grouse infected

dying from it. The strategy for controlling the disease is to reduce contact between ticks and grouse. At sheep-dipping time the graziers pour a chemical down the spine of the sheep so that it eventually covers the whole body, in the same way that dogs are treated for fleas. It is a very effective treatment programme and lasts between eight to twelve weeks.

Beth sheered away from the flames and smoke, and flew low-level down past Grimwith House and across the reservoir.

Floating down to the lowlands, she saw the dry-stone-walled pastureland around the reservoir dotted with sheep. Dropping in among them, like falling leaves, were hundreds of chattering fieldfares and redwings. They had stripped the neighbouring hedges of all their berries and were now intent on another food source – earthworms. They strutted about, stopped, cocked their heads to one side and listened, before making a lightning stab to impale a worm.

Beth's floating approach melded her in with the backdrop of heather moorland and once again, using the dry stone wall as cover, she was able to approach the flock of hungry migrants unseen. Judging it perfectly, she flipped over the wall and, as they rose, stuck out a taloned foot and grabbed a fieldfare with the nonchalant ease of a cricketer taking a catch in the slips.

She carried her kill over the wall and starting plucking the bird – plump and luscious from a month of feasting on berries – under the cover of a clump of rushes.

CHAPTER SIXTEEN

For fourteen days Nidderdale was shut down by continuing sharp frosts. Only the hardy survived. Beth foraged further and further afield, feeding on redwings and fieldfares, and because a big stock of red grouse had been left – too large for them all to survive on the moor in these cruel conditions – weak, sickly grouse were easy prey for her.

Each evening she would drop into one of the three or four roosts she favoured. On this particular night it was bitterly cold and the sky was like lead. She carried out her customary circuits of the roost, checking it out. No other harriers there. She dropped down to her usual couch in the sphagnum moss and rushes. It was deathly quiet, broken for a moment by the blood-curdling scream of a vixen seeking a mate. Beth was soon fast asleep.

She was woken by something soft touching her facial disc. There it was again, but this time on her nape. She looked up. Big white flakes were falling out of the leaden sky. It was snowing. She wasn't going to move as she was quite safe where she was. Head tucked under her wing, she was soon asleep again.

When she woke at dawn it had stopped snowing. Snow lay everywhere: the heather, Great Wolfrey Crag, the rushes and the paths had all been covered during the night. Beth stood up, shook herself, threw up a pellet and took off. As she gained height she could see that all the moorland was blanketed with thick drifts – no use hunting there. Over to the west, though, looking up Wharfedale, there were uncultivated stubble fields and grazing meadows with just a light covering. She would try there. She flew low along the edge of one of the fields, and looking down she could see the tracks of the creatures that had ventured out in search of food – the deep footprints of rabbits, the tiny footprints of long-

tailed field mice and finches venturing out from the hedges, and the bold impressions of winter thrushes and game birds.

Beth had a definite routine. There was less daylight for hunting in winter and she had to make the best use of it. As soon as the first glimmer of sun touched her roost site she flew to the bilberry patch on Great Wolfrey Crag to preen and ready herself for foraging.

One day while she was preening she was rudely interrupted by an unwelcome intruder, a rough-legged buzzard. She shot up from her perch to challenge it, even though it was bigger and bulkier

than her. Yikkering shrilly, she managed to get above the intruder. The buzzard turned on its back presenting eight sets of talons. Beth winged over, neatly side-slipped under it and managed to rake it, dislodging some back coverts. She was much more nimble in the air than the buzzard and, after she had whacked it one more time, it beat a hasty retreat. Nevertheless, yikkering all the time, Beth pursued it relentlessly all the way out of Nidderdale.

Beth returned to her bilberry patch on Great Wolfrey Crag and regained her equilibrium after the skirmish. She was now ten months old, still a juvenile, but capable of breeding. Like her mother, who, a year ago, had sensed changes in her body, she felt the pull of Bowland. Seizing the moment, she left Nidderdale.

'This is something we've found in the past,' says Stephen Murphy. 'Occasionally we find that these birds are presented with two strong pulls. She had this one for Nidderdale where she'd felt safe for the winter and another, stronger one for where she was born. She arrived in Bowland on 2 February 2012.'

CHAPTER SEVENTEEN

As if drawn by a magnet, Beth homed in on the exact area – Mallowdale Pike ridge – where she had been hatched and nurtured by her parents nearly a year before. Fifty feet up, she wheeled purposefully to and fro, searching for a perch, a lookout point, such as the one she had found on Nidderdale. She dived down to inspect a large patch of bilberry, landing on a good, solid stem that she had spotted, then checking all around. In front of her the moor dropped away steeply and gave her an excellent vantage point; behind there was higher ground, so she would not be silhouetted. She did not take these precautions consciously – all was done by instinct.

She left her lookout point on the Mallowdale Pike ridge and dropped down to forage along the stream and pastureland at Whitendale. With the wind behind her, she drifted down the stream, searching for prey. Nothing. When she reached the woods by the aqueduct she hunted along their edges, searching the overgrown ditches. Flap, flap, flap, glide. There! Despite its camouflage, Beth spotted something squatting in a water-filled cow's hoof print. She stuck out a taloned foot and grabbed it just as it flushed upwards – a snipe. She carried it away to eat under the cover of the nearby woods.

For the next four hours Beth foraged continuously, accounting for a short-tailed field vole and a redwing.

Her hunger satisfied, Beth's instincts prompted her to search for a safe roost site. She left the Whitendale area and flew upwards, gaining height rapidly, to survey the moorland beneath her. She remembered she had shared a roost site at Nidderdale with an adult female and her two juveniles. There must be others of her kind around. As the blue gloom of dusk spread over the moor she

decided to keep watch from above and see what happened.

Twenty minutes later she saw two ringtails, either adult females or juvenile hen harriers, drop down into a marshy area at Brim Clough. Even though the light was fast failing, she bided her time and did not follow them impetuously. When she saw a cock bird and another ringtail go in she carried out a couple of low-level sweeps to check the roost. There were plenty of rushes around a sphagnum moss bog and a few small rowan trees – it looked perfect. Down she dropped, just avoiding the roosting male hen harrier, and settled into a warm bed of rushes and moss.

For the rest of the month, whether it was snowing, frosty, misty, overcast or sunny, Beth settled into a fixed routine. She would leave the roost with the others at sunrise, go to the beck and drink, fly up to her lookout point on Mallowdale Pike, and preen and oil her feathers into working order ready for hunting. With the short days in January she had to get her foraging done early. She would then fly back to the communal roost at dusk.

On 2 March she had a frightening experience. She had gone to roost with the other harriers as usual and had been asleep for two or three hours when she was awoken by the distant sound of an approaching six-wheel-drive Argocat. Remembering what had happened at her original Nidderdale roost she let out a yikkering alarm call, alerting the other harriers. Beth followed them as they fled from the roost. The cock bird led them through the dark to another roost site on Whitendale Fell. She was now safe; or at least she thought she was . . .

An article written by Tim Melling, Steve Dudley and Paul Doherty in the September 2008 issue of the journal *British*

Birds sets out very concisely the history of eagle owls on Whitendale Fell in the Forest of Bowland. A single bird wearing jesses – thin straps used by falconers – arrived in October 2005, followed in 2006 by a second bird. They nested and a clutch of eggs was laid but failed to hatch. They persevered, nesting at a different site nearby, and raised three chicks.

Bill Hesketh takes up the story. 'It was 27 July 2007 when Peter Grice, a police volunteer, visited the eagle owl site. After a few hours of watching and no sightings of the owls he went over the fence and, after a while, found the remains of an adult male hen harrier, feathers scattered all over the place. At about the same time, Stephen Murphy recovered the body of an adult hen harrier ringed in Wales.

'On 3 August, a bright and warm day, Bill Murphy and I were on our way home and decided to visit the site. We scanned the area closely but failed to see any owls. We then went over, and about a hundred yards from the nest found the remains of a common gull. We were sure that this was the same kill that Peter Grice had obtained the feathers from, saying that they were the feathers of a male hen harrier. We got in touch with Stephen and told him of our concerns. Mistakes – once in print – are all too difficult to eradicate. In November we took the feathers to the Northern England Raptor Forum Conference to seek the opinion of experts. Paul Irving and others stated that they were definitely from a common gull. Case closed.'

When I visited Bowland in 2012 the two Bills showed me the site, which was now very well publicised and on a

'must-see' list for any bird watcher. They told me that one of the eagle owls had gone as far as attacking a dog being walked along a path about three hundred yards from the nest. It was no wonder that Stephen Murphy was concerned for the safety of Beth, who had chosen on 5 March to roost nearby.

Beth had just killed a lapwing on the pastureland by Whitendale Farm. She dragged it into the cover of a nearby dry stone wall and was just about to start stripping the breast feathers when a strange call – 'hoo-o', repeated several times – made her look up. She cocked her head this way and that until she finally located

the source, a silhouette on the crest of Calf Clough Head. It called again and took off, then, barely flapping its wings, wheeled over Whitendale and disappeared behind Stony Clough Head. It was an awe-inspiring sight: an eagle owl, or 'flying door'. Beth had not seen its like before.

On the following day, just as the light was beginning to fade, she left her foraging area on Dunsop Fell and made her way back to her roost. She had to pass close to the rocky outcrop of Calf Clough Head to reach it. As she did so, an eagle owl, probably the one she had seen the day before, shot out from a ledge and made to attack her. For a second, Beth caught a glimpse of the owl's blazing, blood-orange red eyes and her massive, feather-covered taloned feet ready to strike. Then she was fleeing for her life. Ahead of her was Whitendale Farm. She had the speed but the eagle owl held the key element of surprise.

She thrust down hard, quickly building up speed, flying as low as she dared. Then she suddenly shot up in the air and saw the eagle owl overshoot beneath her, carried on relentlessly by its own impetus. It stopped on the ground, glaring up at her. Beth was safe.

CHAPTER EIGHTEEN

In 2016 naturalist Mike Dilger and a BBC crew arrived at Bowland to film the eagle owls. Bill Hesketh relates what happened: 'The first clutch of eggs laid in January were lifted by collectors. She laid again and raised three young. We took the camera team to a spot we thought would be suitable to film from. They were fantastic, very organised. They had one of those camouflaged hides and they put it up in fifteen minutes. We left them there and walked away as both owls were watching. The crew phoned us later and told us they'd got some fantastic footage. After they'd gone we collected a lot of pellets, soaked them in water, and dissected them to see what they'd been eating. Mostly rabbits – they've made a bit of a comeback on Bowland – a mallard, red grouse and a wigeon, a rare breeder up here. Anyhow, all three young fledged successfully.'

From now on Beth kept away from Whitendale and the eagle owls, and foraged down Dunsop Fell and Burn Fell. The patch of bilberry at Mallowdale Pike was still her lookout spot.

From this vantage point she was able to watch two pairs of hen harriers skydancing. She wasn't quite sure what all the fuss was about. Two weeks later, she was totally confused when a male hen harrier, already paired up, came over and skydanced for her.

'First-year female hen harriers become sexually mature much later in the year than second-year birds,' says Stephen Murphy. 'That's why Beth didn't understand the attention that was being paid to her. She was not flattered, and on 21 March she left Bowland for Nidderdale.'

Why did she move? Normally, once birds return and stay for six weeks or so they are usually faithful to the site for the rest of the season. According to Stephen, she might have moved on for a number of reasons. First, heather burning; many hen harriers are used to it but perhaps Beth just disliked humans and their activities; second, she was not yet sexually mature and interested in the skydancing cock birds; last, the lack of prey available – vole counts were low and meadow pipits had yet to return.

Beth flew from Bowland to Nidderdale in just thirty minutes. As soon as she saw Grimwith House on the edge of the reservoir she gradually lost height and headed for her lookout point on Great Wolfrey Crag escarpment. It was just as she had left it. She settled in on the bilberry patch and preened, fettling her plumage into working order.

She quickly settled into her old routine of foraging, roosting, bathing and preening. The weather began to get warmer and the first meadow pipits began to flutter in. Beth felt a quickening in her body. For the first time, she watched a skydancing male hen harrier with interest – she was ready to find a mate. She started searching, each day scouring a wider and wider area. She finally searched the most northerly part of Nidderdale just below Leyburn. But no luck.

'Beth's behaviour was not conforming to a normal pattern,' says Stephen Murphy. 'We now knew that she was searching for a mate, so what did she do? She left Nidderdale and carried on north to the next grouse moor below Barnard Castle.'

Hen harriers last bred on Barningham Moor ten or fifteen years ago. Beth arrived on the moor on 31 March, having roosted at Rogan's Seat in Swaledale the previous evening.

She flew low as she came up through Holgate Moor and turned to the highest feature, Osmaril Gill, a rocky escarpment dominating the moor. Flying down the valley she saw sheep – Swaledales – everywhere. The grass was thin, bleached brown by the harsh winter. She swung up and settled on a rocky outcrop that gave her a good view of Barningham Moor. There was a pall of smoke drifting across the heather, with a line of men beating the fire, keeping it under control. Beyond was Barningham village, and right in the distance against a busy road was a big lake. A flock of lapwings was displaying around the green meadows on its edges, twisting and turning in the sky above, calling plaintively.

Beth's feathers tightened, her yellow eyes fixed on the lapwings. They were leaving the lake and heading for the grassy uplands below her. Her talons gripped her rock perch more tightly. The flock started dropping down onto the grass and Molinia *pastureland, but the Swaledales took no notice. Soon, all the waders had settled and were busy fossicking for worms and beetles.*

Beth slipped off her perch and turned left-handed down the valley. She knew she was out of the lapwings' sight now. At the last moment she pulled up out of the valley, her primaries brushing the grasses as she flew at low level, using the sheep as cover, towards the flock of feeding birds. Suddenly she was among them. She grabbed one as it rose in front of her and carried it off to her rocky perch on the escarpment.

For the next six days she slipped into her normal routine: leaving her roost, preening, foraging, exploring the moor and roosting. The days were warmer now and the heather moor reverberated with the calls of courting cock grouse marking their territory. Spring migrants – cock wheatears and stonechats – called lustily from heather perches while their mates built their nests nearby.

All this activity was taken in by Beth as a matter of fact; they were all birds she recognised. But then she tensed, leaning forward, her eyes blazing. On an outcrop of rock projecting from the pastureland was a plump, thrush-sized bird she had never seen before. Although similar to a blackbird, what marked it out was its white breast-band. It was a ring ouzel, a rare summer visitor.

Beth took off. Almost immediately she was spotted. The ring ouzel flew up into the air, chattering in alarm, and headed straight for the safety of the big stands of heather nearby. Beth was gaining on her. Then, from out of the blue, Beth was under attack. Screaming with anger, a cock merlin dived at her, striking her on the back. Beth turned to deal with this cheeky little intruder, enabling the ring ouzel to flee to safety.

But Beth was no match for the merlin, who kept screaming and diving at her, until she left the patch of tall heather that he had marked out as his territory.

I knew this area very well as my father had been the rector at Barningham village. His living there was at the bequest of the Milbank family, and they owned Barningham Moor.

In 1975 I was making a film for the BBC on the peregrine falcon and I needed a grouse-shooting sequence. I asked my father to ask Sir Mark if we could film on one of the days he would be shooting driven grouse. No problem at all, he said.

We were blessed with beautiful weather and I had with me a wonderful cameraman, Hugh Maynard. I told him beforehand that the birds would be flying 'like the clappers'. That didn't faze him at all. Hand-holding the camera, he filmed some wonderful shots; one in particular I remember, a grouse crumpling up in mid-air. He managed to pan with the falling bird until it bounced to a halt right in front of him. Hugh was crouching to the right of the butt and he remembers the gun telling him to cross over to the next butt. There were more birds being driven over there and perhaps that gun was a better shot.

There were other shots in the sequence: the driven grouse flying low over the heather, the guns shooting, dogs retrieving and the killed grouse laid out for inspection. One gesture by one of the guns during a drive impressed me. A pair of short-eared owls were put up by the beaters and flew towards one of the butts. The gun in question shouldered his weapon and saluted the owls, acknowledging that they were predators beneficial to man, although he may well have been influenced by the presence of the BBC.

I knew that Barningham Moor had probably only been a heather moor since medieval times, a thousand years ago. Its fauna would have originated during the Mesolithic Period ten thousand years ago, when what is known as the Wildwood – a mix of birch, pine and hazel in its early phase – flourished, covering much of the country. The birds of prey with which we are familiar would have been well established and in better numbers than they are today, and there would have been nearly three thousand pairs of hen harriers. As sea levels rose and fell, and the British Isles became separated from the Continental land mass, the red grouse became distinct from the willow grouse – of which it is a subspecies – in not having white wings.

Aurochs, a huge wild ox, dominated the Wildwood. Franz Vera, a Dutch scientist, proposed that, rather than a continuous canopy, aurochs, munching away, produced a cycle of succession from trees to scrub to grassland and back again. Using the clearings, Mesolithic man hunted the auroch with spears. Wolves and brown bears dominated the ecosystem, which included red and roe deer, badgers, otters and foxes. By the end of the Mesolithic Period the auroch was extinct. Six thousand years ago the Wildwood climaxed with oaks, hazel, alder, limes and elms.

By the Iron Age, three thousand years ago, hunting had given way to farming. On Barningham Moor the early settlers established enclosures whose outlines are revealed during the eight-year annual heather-burning regime. A more permanent reminder of these early settlers is

the 'cup and ring' carvings found on an outcrop of rock within sight of Osmaril Gill. Their exact significance is obscure, possibly signifying a burial place or a site for religious purposes.

Later, a cooler, wetter climate meant that the debris from deforestation did not fully decay, building up and gradually turning into peat. Shade-intolerant plants such as heathers, bog myrtle and cotton grasses found ready purchase in the peat, and from then on grazing and burning turned this perpetually decaying landscape into what we now know as heather moorland, the grouse moors of today.

CHAPTER NINETEEN

Beth flew back to Bowland on 7 April and re-established herself on the bilberry patch on Mallowdale Pike ridge. The need to find a mate – just a spark in her body three weeks ago – was now overwhelming. Nearby, two pairs were firmly established, and even though they were still skydancing, their nest sites had been chosen and prepared.

Beth started skydancing herself, shooting up in the air, corkscrewing as she climbed upwards, climaxing at the crest of her climb for a tip-tilting pause before plunging down, yikkering all the while and pulling out just a few feet from the ground. Again and again she advertised her availability in this way, but to no avail. It was good to be back where she had been bred, but the burning intensity within her body drove her north again to find a mate.

A dramatic landscape – wet, undulating heather moorland – greeted Beth as she flew into Langholm. Flying low, she saw patches of grassland and a few grazing sheep, glens with bubbling burns, dry stone walls holding pasture and edging muddy tracks, and geometric strips of conifer plantations. She swung upwards to inspect the heather moorland. To get there she had to pass a massive stone structure, the Malcolm Memorial overlooking Langholm town, which was being admired by a group of walkers. She jinked to one side and continued up to the grouse moor, where she was rewarded by the sight of three pairs of hen harriers skydancing. Sadly there was no spare male looking for a mate so she sped on, eager to continue her search on the grouse moors further north.

Langholm is best known as the 7,600-hectare grouse-moor estate belonging to the Duke of Buccleugh, where

a project has been running since 1992 to establish whether a grouse moor with breeding raptors is a viable proposition. This was at the behest of grouse-moor owners, who were becoming increasingly vociferous in demanding an end to the hen harrier's ever-increasing population.

The prime movers behind the five-year project were two bodies: the Centre for Ecology and Hydrology, and the Game Conservancy Trust. Other interested parties in the study included English Nature, the RSPB, the Moorland Association and the National Association of Gamekeepers.

The late Dr Roger Clarke was an accountant who worked as an observer at Langholm. He was originally a fanatical fisherman and moved to Reach in Cambridgeshire for the fishing. On Adventurers Fen he met that wonderful bird artist, Eric Ennion, who introduced him to marsh harriers and hen harriers. These birds transformed Roger's life and he got rid of his fishing tackle, which is how this very shy genius came to be part of the team watching hen harriers at Langholm.

A subsidiary project at Langholm was launched between 1998 and 1999 to discover whether predation of grouse chicks by hen harriers could be reduced by diversionary feeding. The use of this method was to be one of the key factors in the proposed plans to save the English hen harrier. Roger Clarke was a volunteer at Langholm during that period.

Diversionary feeding was not a new idea; in 1989 John Phillips and Robert Boutflower had suggested using a dovecote on a grouse moor from which pigeons would

be released to divert peregrine falcons away from taking grouse, and the following year the Hawk and Owl Trust was invited by the Joseph Nickerson Reconciliation Trust to take part in a diversionary feeding project to reduce the impact of hen harriers on red grouse. It was said that at three hen harrier nests monitored in Perthshire, 550 grouse chicks were brought in by the adult harriers to feed their young.

In this case it was suggested that open-topped pens should be set up about 250 yards from the harrier nest site, with either live chicks or rodents being used as diversionary prey. Unfortunately lack of funding meant that the project never got off the ground.

Meanwhile, the results of the trials at Langholm between 1998 and 1999 were astonishing. Predation of grouse chicks was reduced by 86 per cent when the parent birds were fed day-old chicks over the critical six-week period. Here surely was a way forward. But the grouse-moor owners suggested that the practice would attract more predators, and they had issues with the practicalities and long-term consequences of the project.

'The chicks and rats were placed on T-posts about nine yards from the nests,' says Roger Clarke. 'It was clear that the harriers much preferred the chicks to the rats; during one July hide watch I witnessed the female pick a rat off the T-post in flight without pause, fly with it across the burn to the other side of the valley and drop it randomly from a height as she flew steadily on – not a cacheing act. However, some rats were fed to the young. On another

occasion I watched a female on the nest break into a rat at the back of the neck and peel down the skin to dig into the flesh. One tally-up showed that chicks and rats supplied in the ratio of 4:1 were consumed in a ratio of 17.4:1.'

My contact at Langholm was Simon Lester, who became head keeper there in 2008. Before that – which is how I knew him – he was head keeper at Holkham in Norfolk. Whenever I rang him and asked how the hen harriers were doing he would reply, 'The hen harriers are fine, no problem. It's the buzzards I'm worried about – there are just too many of them!'

So what has been done to stop the age-old conflict between conservationists and grouse-moor owners? In 2006 English Nature set up the Hen Harrier Stakeholders Group Committee. The stakeholders were made up of conservationists, grouse-moor owners, scientists and legislators, each with their own particular goals. The Environment Council was brought in to try to resolve the conflict between the interested parties, particularly between the conservationists and the grouse-moor owners, and the press had a field day misinterpreting scientific findings. Conservationists pointed out that there was suitable habitat for three hundred-plus pairs of hen harriers on England's heather moorland. There were discussions on brood management, and although there were many meetings they always ended in stalemate and the group finally disbanded.

Stopping only to roost and forage Beth flew on searching the grouse moors for a mate. By 19 April she was soaring over the grouse moors around the Dornoch Firth, north of Inverness.

I knew that area very well. In 1975 I had filmed the late Stephen Frank, one of the world's foremost falconers, flying his trained peregrine falcons there. He lived in a remote cottage on the edge of the moorland behind Dornoch, and had also bred a world-renowned line of pointer dogs. Stephen, his falcons and his pointers formed a unique partnership, and he had access to several local grouse moors for hawking.

The pointers did all the initial work, ranging upwind over the moor to seek out grouse. As soon as they found them they went absolutely rigid – head thrust forward, one front paw raised, tail straight out behind. Now Stephen unhooded his falcon and cast it off, letting it climb into position above the crouching grouse. Meanwhile Stephen ran around, watching the circling falcon all the time, until he was upwind of the grouse. A good falcon such as Stephen's would 'wait on' at near 1,000 feet. As the falcon turned downwind, Stephen rushed in and flushed the grouse. The falcon fell like stone, reaching a speed of 150 mph, the air whistling through its primaries sounding like tearing linen. Just before it reached the grouse, skimming along above the heather, the falcon pulled out from the vertical. Its outstretched taloned feet hit the grouse a glancing blow, knocking it over into the heather. It was a good team effort. In an interesting footnote, Stephen told

me that during the hawking season he wore out a pair of gym shoes every fortnight just by running through the heather.

Back to the present, and I like to believe that Beth suddenly became indecisive, pulled in two different directions. Abandoning her search for a mate, she turned back above Stephen's cottage and returned to Bowland.

CHAPTER TWENTY

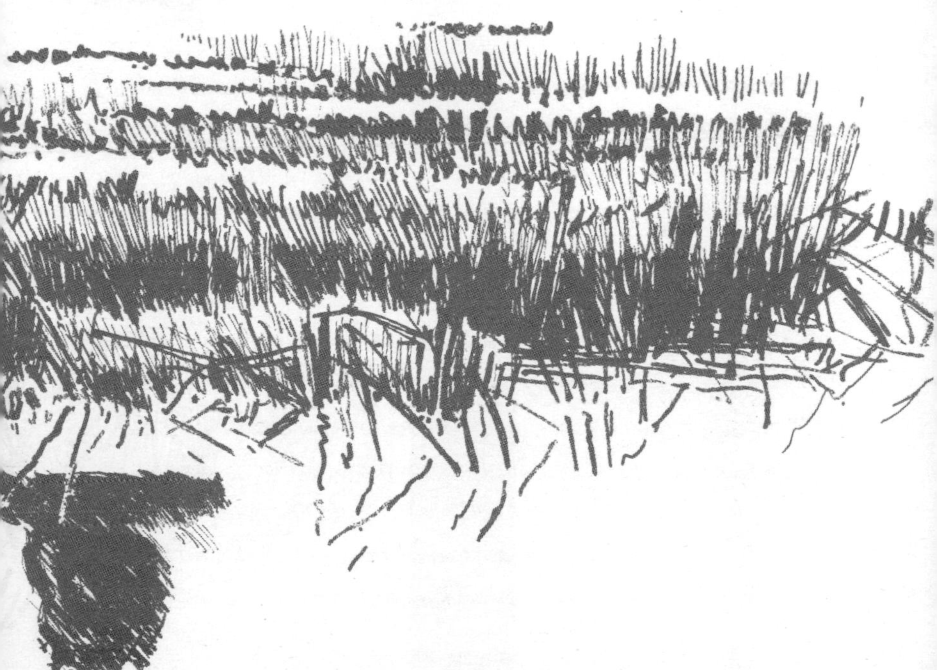

The warmth of the sun was on her back as Beth floated down on to Bowland searching eagerly for any skydancing males. She was not to know that this was to be a lean breeding season for hen harriers in England, with only one successful nest at Bowland raising four young. Beth wheeled round, flapped her wings, braked with her spread tail and reoccupied her lookout perch on the Mallowdale Pike escarpment.

Her bilberry patch was now sporting new green shoots, as were the stands of heather surrounding her. Five yards away, a cock stonechat – a sparrow-sized bird – suddenly appeared on a stand of heather to boldly proclaim his territory. He was a handsome bird, with a black head and throat, white half collar, dark upper parts, white rump and chestnut undersides. His tail flicked up and down in time with his click-clacking call, while deep in the heather his sober-plumaged mate was incubating five eggs in her beautifully fashioned nest of mosses and grasses. The eggs were due to hatch any day now. All around her other birds were busily going about the yearly ritual of courting, nesting, egg-laying, incubating and raising young.

Beth flew off, dropping down from the escarpment to forage. As she flew low over the old Roman road she started to inspect the heather bordering Salter Fell. Hunting with the wind, wings held high, she quickly covered the heather. She turned into the wind and followed a beck bubbling down the fell. She was searching, listening for the give-away rustle of prey in the new growth of Molinia *grasses on either side of the beck. Back and forth she went.*

She was about to give up and return to forage the heather when two shapes hidden by overhanging grasses slid into view. She flapped her wings, spread her tail, pirouetted and dropped

down, grabbing into the grasses. It was two leverets in their 'form'. A blood-curdling scream from the leveret as it was carried away alerted the doe. She raced after Beth, who was flying low, weighed down by the leveret. As she caught up with her she jumped in the air and lashed out with her hind legs, catching Beth across her breast-bone. The leveret let out one last, desperate scream, strangled at the death as Beth gripped tighter, and the doe attempted a final, despairing boxing jump at Beth before she disappeared and returned to care for her remaining leveret.

Beth flew on, searching for a deep stand of heather where she could eat the leveret in peace.

Half an hour later, Beth returned to the nearby beck to drink and to clean her beak, breast feathers and talons. Her crop bulged satisfactorily, an indication of a good meal taken. She flew back leisurely to her lookout point on Mallowdale Pike escarpment and gave her battered plumage a good preening.

Later, as the shadows lengthened in the blue light of dusk, she dropped down to roost at Brim Clough. She was the only harrier there.

In the morning she cast up her pellet and continued her search for a mate.

Roger Clarke was acknowledged as one of the world's most expert analysers of the pellets of birds of prey, receiving his doctorate from the University of Liverpool in 1999 for his thesis titled 'Aspects of raptor, Falconiformes, feeding ecology: an approach using pellet analysis'. Between 2004 and 2005 Stephen Murphy took him to a couple of roost sites on Bowland to collect hen harrier pellets.

'There he was in a green BMW 325, early in the

morning, taking me off pellet collecting,' recalls Stephen. 'What is this guy about? I wondered. We went to a roost where we knew through radio tracking that there were birds arriving, and on the first day we collected a few pellets. That evening we watched the birds coming into the roost sites. The following morning we collected a load of pellets – seventy pellets from one, and about forty from the other. Roger was very knowledgeable, but very humble. I remember opening a pellet and looking at all these different carapaces and mandibles. From about six feet away he spotted a goldfinch's beak. And then he started on the small mammals. He could identify each one by its micro-teeth, little bones and ribcages. It was a real privilege to watch him at work.

'Eventually, he sent me his analysis of all the pellets. It was amazing. Most of the female hen harriers on Bowland were eating grouse plus the occasional leveret. The males were eating everything – common snipe, wrens, goldfinches, linnets, meadow pipits and redwings. To see a hen harrier catching a redwing is amazing. It's a mirror image of a sparrowhawk's technique – air brakes on, put out a taloned foot and grab one. Real acrobatics.'

Shortly after the Hawk and Owl Reserve at Sculthorpe Moor was opened, Roger, who was our scientific officer and accountant, made an appeal for volunteers to help monitor hen harriers that were flying in to a winter roost from October through to March. He organised a meeting in an upstairs room in a nearby pub and gave us a lecture on hen harriers at roost sites. He told us that most of the birds we

would see would be ringtails, which were either immature males or females or adult females, and that if we were lucky we might see a beautiful silver-grey male hen harrier.

It was about 3 pm when we arrived at the roost site. Roger led us to a good watch point and, binoculars poised, we settled down to wait. There was plenty of time for a good look round. We were sitting on a hill looking out over the common, which was enclosed by silver birch to the east and by a mixture of silver birch and conifers to the south. There was a copse to the north where we had parked, and it was open behind us. The roost itself was largely covered by heather, and in its centre was a damp, paler area dominated by *Molinia* grass and sphagnum moss. The whole area was dotted with silver birch seedlings and there were one or two stands of gorse.

Roger directed us to watch the skyline above the silver birch to the east, the area from which the harriers would appear. Sure enough, about ten minutes later a ringtail appeared and started flying to and fro along the treeline, checking that the roost was safe. In no time at all it was joined by four others.

Reassured, one by one the harriers dropped down to fly backwards and forwards, lower and lower, over the pale, damp area. Although the light was dying one could still follow the action, and now and then a harrier would drop into its familiar roost site in the sphagnum moss. One time there was a commotion when a harrier dropped into a roost that was already occupied and both birds sprang in the air. There was a bit of a kerfuffle before the offender,

probably a newcomer, went off to settle elsewhere. It was the first of many visits that I made to this winter roost site.

In 2003 Roger published a paper with the title 'Monitoring the diets of farmland winter seed-eaters through raptor pellet analysis'. Any hen harrier enthusiast in the eastern counties of England will have seen winter-migrant harriers foraging along the fringes of sugar beet fields. Their slow flap, flap, flap, glide flight enables them to pounce on any unwary finch feeding on the prolific amount of seeds produced by weeds like the fat-hen plant.

Through analysis of pellets found at three East Anglian harrier roost sites Roger was able to show that five passerines – skylarks, hedge sparrows, greenfinches, linnets and reed buntings – were extremely dependent on the narrow range of annual weeds found on the edges of arable crops. Roger's paper, written in collaboration with Pete Combridge and Nigel Middleton, went on to emphasise that farmers should try to nurture populations of common weeds in arable crops and not fall back on the easy fix of genetically modified, herbicide-tolerant crops that would put added survival pressure on a number of farmland birds.

Just after that first visit to our winter roost site, Roger and Stephen Murphy set about trying to catch hen harriers at winter roosts so that they could radio-tag them. 'Roger knew that there were three or four harriers coming in to roost at our site every evening,' says Stephen. 'We had the radio tags – we just needed to catch some harriers. We were going to use two lightweight fishing poles with very fine mesh strung in between. We arrived at the site

and spotted where the harriers went in to roost, an area about the size of a tennis court. There were four of us, and the plan was that three would position ourselves upwind of the site, then move forward to the roost holding the poles upright with the fine mesh stretched between. The remaining member of the party would go downwind of the site and on an agreed signal would rush in to flush the harriers into the net.

'We waited for the light to fade, then moved on to the roost site, stumbling about in the heather and sphagnum moss until Roger signalled that we were in position and ready. As planned, the fourth member of our party rushed forward, clapping his hands, and the harriers headed straight for the net just like bullets. Ten feet away, even though it was dark, they saw the net, shot up vertically over it, dropped down again to low level and sped away. I was amazed that they could see so well in the dark. We sat there bemused. What next?'

Over the next eight days Beth ranged widely over Bowland. She seemed unsettled, not necessarily using the same roost site each evening. RSPB staff saw her ripping up bits of heather and skydancing.

Beth was restless, her ancestral memory pulling her north again. Maybe her original roots were to be found with the hen harriers on Orkney, which was still their stronghold.

'Of one thing I'm sure. I'm never going to try and second-guess where Beth is off to next,' says Stephen. 'On 3 May

Beth left Bowland and sped north. Thanks to the satellite tag we could track her movements exactly. By the 4th she had reached a grouse moor at Biggar, south-west of Edinburgh, and on the 5th she roosted by Loch Bruicheach about fifteen miles west of Inverness. I knew this area well. I've watched hen harriers passing food on the plantation rides above Drumnadrochit and it's the haunt of Slavonian grebes on the lochans there. On the loch there are black-throated divers and the occasional greenshank.

'On 6 May Beth left her roost at 4.47 am and sped north, arriving over Forsinard at 11.39 that morning. She didn't stop there but carried straight on.'

I was sad about that. I'd made a film at Forsinard called *The Last Great Wilderness.* It was part of a series and Bruce Pearson, a long-time friend and illustrator, was the programme's presenter. This is his description of the landscape:

> Ten thousand years ago, when the glaciers of the last ice age finally retreated, the landscape left behind was a waterlogged, treeless, tundra-like wilderness. In time, a deep mantle of peat developed, a vast saturated blanket across the Flow Country. Its surface was broken by a myriad of small pools on the surface of the peat. The cool, wet and windy climate of these parts and the acidity of the water encouraged a complex ecosystem. In acidic water with little oxygen the normal process of decomposition virtually ceases. So, for thousands of years, the continued process of decaying plant material forms peat over six metres thick. Part of this ecosystem

are the sphagnum mosses that form a deep, spongy layer on the surface. The sphagnum is the habitat of a rich patchwork of lichens, sedges and special spectacular plants like the sundews, which catch insects on their sticky tentacles.

At Forsinard the blanket bog has reached its greatest expanse. It is an extremely rare habitat: not only is it the largest single area in Britain but also the largest expanse of blanket bog in the world.

How Bruce managed to set up his easel and paint there God only knows. To me it was like walking on a trampoline – very tricky.

While we were there we filmed dunlin, greenshank, curlew, golden plover and common snipe. There were red-throated divers on the lochs and, of course, there were hen harriers. I remember watching a dazzling silver cock bird hunting along the edge of a conifer plantation – why hadn't Beth stopped off there?

But she had instead flown on for another hour and a half before reaching a grouse moor at Loch Shurrery, about six miles south of Thurso in Caithness, having flown 120 miles in just under eight and a half hours. On 8 May she headed south again, unerringly homing in on grouse moors, and was tracked at well over 4,000 feet as she passed the summit of Ben Macdui in the Cairngorms during her journey south. Unlike her dash north, Beth positively loitered on her return journey, arriving back at Bowland on 22 May.

Her return coincided with Eddie Anderson and myself arriving at Bowland for talks with Stephen Murphy, Natural England and United Utilities, who sponsored the RSPB presence on Bowland, Phil Gunning, who rented the grouse moor for walked-up shooting only, and Jude Lane, the RSPB lead. Later in the day, Stephen took us up onto the moor to show us the general areas where the hen harriers bred, foraged and roosted.

On 23 May we met the two Bills – Hesketh and Murphy – who took us out and showed us where the eagle owls nested and how they had come to be there. Beth was somewhere around, but we never saw her.

Beth was foraging on Burn Fell. A familiar routine – flap, flap, flap, glide – as she searched upwind. Her shadow played on the lush grass and sedges as they rippled in the breeze, her eyes and her ears behind her owl-like facial disc focused on any movement or sound that might pinpoint a vole's position.

She turned at the dry stone wall to hunt downwind, holding her wings in a V-shape to let the wind take her down the fell. She knew that there were plenty of voles – she'd caught two on Burn Fell only the previous day – but perhaps the one successful pair of hen harriers had been back and scoffed the lot.

With her acute hearing she identified a rustle in the grasses ahead. Her ears triangulated the exact position of the sound, then she spread her tail, flapped her wings to slow down and hovered directly overhead. She saw a movement and plunged down, her taloned feet perfectly aligned with her eyes. The vole had no chance.

She continued foraging over Burn Fell for the next two hours and caught three more voles before returning to her lookout point on the rocky outcrop at Mallowdale Pike. As she preened she noticed that the stonechats, who were nesting in the tall stand of heather nearby, had reared five young. They were perched in the heather, calling for food. Both parents took it in turn to feed them an assortment of beetles, larvae, moths and spiders. The hen bird would soon return to her nest to raise a second brood.

All around her, young were being raised. She was dispirited. Why couldn't she find a mate? With that thought festering in her mind she flew off to roost alone at Burn Fell.

It was pouring with rain on the morning of 25 May as Beth left her roost. She went to her bilberry patch lookout point on Mallowdale Pike escarpment, preened and without any hesitation, flew off into the gathering storm towards Nidderdale.

'At about this time I began to be worried about Beth,' says Stephen Murphy. 'Where was she? So, on a hunch, I asked Mick Carroll to go to Nidderdale and see if he could locate her. He phoned later and said he'd seen her in appalling weather, "going like a bat out of hell" towards Grimwith Reservoir.'

CHAPTER TWENTY-ONE

Through the blue-black gloom of the driving rain, occasionally riven by lightning, Bowland Beth at last glimpsed Nidderdale, the moor that had been home to her after she had fledged. As the thunder echoed over Wharfedale, she realised this might be her last chance to find a mate. But first she needed to roost up for the night. On the edge of the reservoir she saw two figures peering up at her. Man.

She jinked to one side, fearful of their silhouettes, then swept out over the reservoir and soared up above the moor till she saw one of her old roosting sites, Hard Gate Moss. There were the familiar clumps of purple heather and the inviting sanctuary of rushes, grasses and bog. She flew past, turned into the wind and flapped slowly over the area. Perfect, just as she remembered it, down to the tiny beck trickling through the bog. She flapped her wings, spread her tail to land gently on the soft bed of sphagnum, then settled down between two hummocks, a comfortable roosting place, enclosing and warm. Within minutes she was fast asleep.

She was awake at first light and flew over to the beck. Perching on a craggy boulder, she slaked her thirst, and once she was satisfied she started to preen. First she rubbed her head and beak against the oil gland at the base of her tail, then used her beak to comb through every one of her wing feathers to ensure that they were flightworthy. Her tail feathers received the same meticulous attention. She puffed herself up, and after a good shake cast up a pellet, the undigested bones, feather and fur of yesterday's prey. The sun was now up, promising a warm and dry day, and she was ready to hunt.

Flap, flap, flap, glide. She flew down the beck, which bubbled vigorously as it flowed downhill. The banks on either side were edged with grasses that melded into the heather, and here and there she noticed small piles of nibbled grass shoots and holes in the bank – water voles. She cut her speed. There! She plunged down, grabbing the vole before it could reach the safety of its home.

She mantled over the dead creature, killed instantly by her needle-sharp talons. Her beak ripped into it, butchering the carcass into bite-sized chunks. Much bigger than a short-tailed field vole, it made a satisfying start to the day.

She then took off, ringing rapidly up in wide arcs to gain an overview of the moor. She could see the familiar lines of grouse butts and jinked off to one side. Further north, on another moor, she could see a ridge thick with heather. She flew towards it, circled overhead and surveyed the scene. Perfect. She closed her wings and dived down, then flapped her wings, spread her tail and flew low along the ridge searching for the ideal place to nest. There it was, a shallow gully with tall heather and an outlook onto the moor. She landed, looked around and started pulling at the grasses, throwing them in the air. She shot up into the sky and threw herself into her all-too-familiar skydance routine, then flopped down, panting, and perched on a boulder to scan the moor and the sky above. Nothing.

For three days she set up her stall at her chosen site and religiously skydanced to attract a mate, but her only onlooker was a raven. He set a beady eye on her, uttering a contemptuous 'kronk-k' as he flapped unhurriedly past.

On the fourth day, catching her breath after yet another awe-inspiring skydance – and just as she was thinking she must move on – the unbelievable happened.

At first all she saw was a speck in the sky. But as it grew closer it resolved into a male hen harrier. Was she going to be

disappointed again? She flopped down into her chosen site and started tearing at the grasses, throwing them in the air, then watched anxiously. He altered course. She flapped up onto her perch just as he dived towards her.

Determined to impress her, he went pell-mell into his own skydance routine. Yikkering, he skimmed overhead, making her duck, before shooting up into the sky. He flipped over onto his back before rocketing up again. As he levelled out, three hundred feet above her, his blue-grey plumage merged with the blue of the sky. He tipped over and plunged in a death-defying dive towards her again. She was entranced. At last she had a mate.

Two hundred yards away a black silhouette broke the outline of a dry stone wall. There was a brief glint of sun on metal. It steadied.

Bowland Beth never knew what hit her. One moment she was spellbound by the skydancing, a fraction of a second later she found herself lying in the heather, her left leg broken and bleeding, and six of her tail feathers missing. In great pain she managed to right herself and took off, trailing a plume of blood. She cleared the ridge and planed down across the moor. Perhaps she'd be safe on one of the other moors. Flap, flap, flap, glide. Her vision dimmed. Flap, flap . . .

She crashed into the heather.

The cock bird landed next to her. What was wrong with her? He gave her a gentle prod with his beak. She turned her head towards him. She gasped in pain, a shudder shook her body and her bright yellow eyes sealed over. The pain left her and she flew to the Elysian Fields, to a grouse moor where there was a natural world of interlocking ecosystems and where birds of prey were never persecuted.

'Another fix from Beth's transmitter on 11 June showed that she had contracted her foraging range to the grouse moors around Nidderdale and Colsterdale,' says Stephen Murphy. 'I was sure she had found a mate.'

For three days Beth's mate returned to soar round her final resting place. On the third day their bond was broken by blinding rain, great scuds of it sweeping over the moor.

'It was the start of one of the wettest Junes in living memory,' recalls Stephen. 'Heavy cloud cover meant that for several days we couldn't get an accurate fix on her. On about 14 June I became really concerned. Maybe the transmitter had failed. I contacted the manufacturers and asked whether the last fixes were reliable. They confirmed my suspicions that Beth had died sometime between 8 and 11 June. I was therefore able to plot her approximate position on a map. I contacted the landowner. He couldn't have been more helpful – so cooperative. He arranged for the head keeper to help me search.

'Using a hand-held scanner I managed to locate Beth at 11 am on 5 July. She was lying face down in a patch of heather and bilberry. The satellite tag was plainly visible. Six of her tail feathers had been cut through. A post-mortem showed that she had probably been shot. One of her legs was broken and her femoral artery had been nicked. She probably would have been able to fly a few miles before she bled out and collapsed onto the grouse moor where she was found. Later, cutting-edge scientific analysis showed fragments of a lead pellet or bullet embedded in the leg bone. She had definitely been shot.

'Beth was a beautiful bird – an amazing bird – and I feel so privileged to be the only human to have held her while she was just a bundle of earth-bound feathers and attitude. Her story is remarkable. We should be celebrating her life now and her becoming a parent, and tracking her sons and daughters.'

We will probably never know exactly what happened.

Perhaps this fearless, naive bird went a wingbeat too far and had to run the gauntlet to regain the grouse moor that she knew as home. We grieve that, illegally, she was cut down in the prime of life. I hope she has not died in vain.

EPILOGUE

Three years later, in September 2015, Eddie Anderson and I returned to Bowland to talk to Stephen Murphy about what effect Beth's death has had on the fortunes of hen harriers in England. I told him that Henry Williamson, my mentor in my early days of wildlife film making, always used to quote Thomas Hardy: 'If way to the Better there be, it exacts a full look at the Worst.'

'Beth's death created huge publicity,' says Stephen. 'There was widespread public condemnation from within every conservation body. But worse was to come. A year later, in 2013, the hen harrier had disappeared as a breeding bird in England. What I couldn't take in was that, year after year, between 2003 and 2008, hen harriers had bred very successfully at Bowland, raising 138 young. But from then on it was all downhill. What had happened?'

About ten or fifteen years ago some – and I emphasise, some – moorland landowners and shooting-syndicate managers realised that the red grouse was a cash crop that they could cash in on, given the minority group of very rich people who were prepared to pay £3,000 to £4,000 a day to shoot driven grouse. Medicated grit had got rid of *strongylosis* and more effective sheep dips put paid to looping ill, the two diseases that had previously badly affected the red grouse, causing 'crash' years in the population. And any predator, four-legged or winged, was mercilessly eradicated. Top of the list was the hen harrier – it is no secret that hen harriers do eat grouse chicks.

In June 2013 the RSPB were filming a nesting hen harrier in Morayshire. To their amazement a keeper walked

into the frame, flushed the hen harrier from its nest, raised his shotgun, two shots were heard and the harrier killed. Cool as a cucumber he collected the dead bird and walked off. The keeper was charged and taken to court.

Stephen offered to take us to Nidderdale, where Beth had spent the last five months of 2011 and the first two months of 2012. We parked with a grandstand view across the Grimwith Reservoir to the great mound of Nidderdale, trisected by three streams emptying into the reservoir. High up on the heather moorland we could see Great Wolfrey Crag, the escarpment of exposed grey rock that would have been a landmark for Beth on her flight from Bowland.

'A lot of grouse estates now have incredibly high numbers of grouse,' says the RSPB's Guy Shorrock. 'Dr Peter Hudson of the Game and Wildlife Conservation Trust suggested that to have sustainable driven grouse shooting you needed about sixty grouse per hundred hectares. But the numbers on some of these estates in the north of England are incredible now. I think it's over two hundred birds per hundred hectares. It's like factory farming. And there's a bit of an arms war going on too. Keepers are developing new methods to control birds of prey, more efficient methods to draw in males close enough to shoot them when they've been out foraging. Some of it is quite blatant.'

For instance, in 2016 two bird watchers walking along a moorland ridge overlooking a valley in the Peak District were delighted to see a male hen harrier perched on a fence post. A quick scan through binoculars showed,

however, that it was a replica of a male hen harrier perched on a pole. A more thorough search of the valley bottom revealed a man in a camouflage smock partially concealed in the vegetation nearby, his Land Rover parked five hundred yards behind him. When he suddenly became aware that he was being filmed, he jumped up, dashed forward, gun in hand, gathered up his replica hen harrier and scarpered.

Worse still was the behaviour of a young keeper in north Yorkshire who was photographed checking three pole traps, which are not only cruel but quite illegal. Such blatant behaviour is quite rare, and the persecution is usually much more discreet. Gamekeepers now have six-wheel-drive Argocats and quad bikes, meaning there is no part of a moor that a keeper cannot reach quickly and, for example, chill a clutch of hen harrier eggs with a handful of ice cubes and be off the moor in a flash.

The modern keeper is kitted out with state-of-the-art night-vision goggles, together with 'scopes and thermal-imaging equipment. When the latter is used at a harrier roost at night it projects an image of a hen harrier glowing like a Belisha beacon, and the birds are easily taken out. Keepers are rewarded handsomely and the tips they receive from guns at the end of a day spent shooting driven grouse are lavish. It is no wonder that they are motivated to kill birds of prey illegally.

Now, as I have already discussed, red grouse have been hit by a new disease, bulgy eye syndrome or *Cryptosporidium baileyi,* as it is scientifically named. A Game and Wildlife

Conservation Trust statement in 2014 explained that the disease, which first appeared on grouse moors in Northumberland in 2010, 'infects the bird's sinuses, causing swollen eyes and an excessive production of mucous; it is like a particularly nasty head cold.' By 2014 it had taken hold on the grouse moors of the north Pennines.

At a recent conference in November 2015 the same Trust announced that the disease is caused by a protozoan parasite. The build-up of mucous in the trachea and larynx can be so severe that infected birds 'gargle' as they take off, and the cock birds have difficulty in calling when they land. Cross-infection occurs through the transfer of oocysts released by the host parasite, with driven grouse shooting, in which different family groups of grouse are flushed and mixed in flight, being one way in which the disease spreads. Other ways discussed at the conference were via the piles of medicated grit that grouse visit each day to aid their digestion, and by human feet and the tyres of vehicles.

It is quite obvious to most people that a high-density population of red grouse plus Cryptospiridium is a disaster entirely of the grouse-moor owners' making, particularly allied to the fact that their zero tolerance towards raptors means there are no longer any hen harriers to pick off the weak and sickly grouse, which then continue to spread the disease.

After the 'black' year of 2013, with the hen harrier extinct as a breeding bird in England, Defra sprang into action and reconvened the Stakeholders Group. A six-point action plan was proposed that included trials of

lowland reintroduction of hen harriers in England and a brood-management scheme.

I felt that the Hawk and Owl Trust should be doing much more to help the hen harrier. I suggested to our chairman, Philip Merricks, that he look into the feasibility of undertaking trials of brood management and the reintroduction of hen harriers to lowland areas. One of our reserves, Fylingdales Moor, had once been a driven grouse moor. It was guarded by RAF Fylingdales and there was a secure area where a trial could take place.

Philip arranged meetings with the interested parties. A scientific committee was formed consisting of Professor Ian Newton, Professor Stephen Redpath and Des Thompson of Scottish Natural Heritage. Ian Newton's feelings when I spoke to him were that trials on brood management should be given much more thought, and although there wasn't a large enough population at the moment to justify it, there was nothing against re-introduction of birds from abroad to lowland areas such as Salisbury Plain or Exmoor.

Unfortunately the plans for brood management were 'leaked', and when Mark Avery, the former conservation director at the RSPB, got wind of them the term 'brood meddling' was born. The Hawk and Owl Trust became reviled by conservation groups and we lost our president, Chris Packham. The attitude of those on the ground looking after harriers was exemplified by Bill Hesketh: 'It's like saying to a burglar, "There are ten houses – you can burgle five and we'll overlook what happens to the other five."'

No English hen harriers bred in 2013. It was extinct. But in 2014 there was a recovery with four breeding pairs in England. Two of the nests were at Bowland, one in the Peak District and another up in Cumbria. Seven birds were satellite-tagged. Only two birds survived, both from Bowland. But there was better news from Wales. Steve Thornton, who had worked on the reintroduction of red kites to Rockingham Forest, rang to tell me that he had discovered a new hen harrier nest in Wales.

'It's the most wonderful site,' said Steve. 'I was looking down into a basin with big hills all around. There's heather moorland at the top, then some forestry. Lower down the heather gets sparser and sparser. At the bottom it's damp with rushes and tussocky grass. There are a couple of willow trees and patches of broom and gorse.

'I heard a bit of "wittering" and saw the male hen harrier float down into the valley. Almost immediately the female flew up, underneath and behind him. The minute he dropped his prey, she turned on her back, stuck out a foot and caught it. I watched her fly down and land on a big boulder. From there she flew to a line of fence posts. She moved along the fence, hopping two posts at a time, until she dropped in behind a sparse patch of gorse.

'Meanwhile, the male had found a thermal and was floating up from the valley bottom. As soon as he was level with the high ground he started hunting again. Altogether I saw five food passes, so the female was obviously on eggs. I was lucky with the weather – real barbecue weather – and it was superb watching that silver-grey male floating around.'

Steve's news was a shaft of sunlight breaking the pervading gloom in England regarding the fate of the hen harrier. Mark Avery was appalled at the continued illegal killing of hen harriers and organised a petition to have driven grouse shooting banned. In just ten months he managed to gain 22,000 signatures backing his petition. He and Chris Packham supported the 'Hen Harrier Day' rally in the Peak District, which, despite the appalling weather, was very well attended and is now an annual event.

The RSPB reported that 2015 was the most succesful breeding season for hen harriers in England since 2010 – eighteen young were successfully fledged, although it could have been much better as all six nests at Bowland failed. Were there any mitigating circumstances? It had been a bad vole year and March and April had been cold, so it was not good for grouse either. But the disappearance of four male hen harriers was suspicious. I assumed that they had been shot while out hunting.

'The trouble is that for a cock bird going out hunting there's only one way out of Croasdale at Bowland,' says Terry Pickford. 'It's therefore easy for a keeper to position himself on their flight path and shoot the birds as they are going out or on their way back. It highlights a clear and unequivocal message from the game-shooting industry – these birds are not welcome on grouse moors.'

Later that same year a juvenile female hen harrier named Annie fledged at Langholm in the Scottish Borders and, fitted with a satellite tag, was tracked by Stephen Murphy and Natural England. She was later found dead on a grouse moor in South Lanarkshire. She had been shot and because she had not been killed cleanly, had struggled away like Beth, perhaps surviving for a few days before dying. This is why the shooter had not been able to collect her body.

Stopping persecution is the key to the hen harrier's survival as a breeding bird in England. People tend to forget that since 1954 this bird has been fully protected

by law under the Protection of Birds Act. We need more wildlife crime officers and, above all, lighter satellite tags on harriers, working through broadband, which will give an instant fix, twenty-four hours a day, to a mobile phone.

Stephen Murphy pulls up alongside a dry stone wall that overlooks a wet, rushy meadow. 'This is a new site,' he says, 'a saucer-shaped depression at 220 metres above sea level. It's too wet to graze, too wet for grouse. It's a six- or seven-acre patch of land that's gone back to a natural state, with lots of rushes and a small patch of willows. I've had a word with the farmer and he's aware that it's a harrier roost. We need more passionate birds of prey watchers – like Bill Hesketh and Bill Murphy – to monitor roost sites so that we know the numbers of harriers and where they are.

'We rely very heavily on the amazing work done by Natural England volunteers. A couple who have recently passed away and are sadly missed are Mick Carroll and Ted Parker. But at present, Derek, Pete, Mike, Gavin and Pat volunteer their time and expertise for the hen harrier project and go out in all weathers to monitor our uplands for this amazing species.

'We also work closely with members of the North of England Raptor Forum for information. The area is split into eight different sectors and the sightings they gather filter down to me. I've got one old chap, Billy, who sits on a disused railway bridge every Sunday watching for harriers. He's been doing it for twenty years and his records are amazing. Further down the old rail network there's another guy, Ken, who sits in his car and watches. Some days I get a sighting off Billy to the north and then within ten minutes I get a sighting from Ken. It'll be the same bird, but that doesn't matter. Long hours in the field – there's no substitute for that. It's the way we chaperone our harriers.'

We press on, as Stephen is anxious to show me Beth's final resting place. We climb up onto the heather moorland and soon come across an area of heather being burnt, a great pall of smoke and a leaping snake of flaming heather being swatted by a line of keepers. 'Heather burning on its present scale has got to be reined in,' says Stephen. 'It finishes on 15 April – that's too late, and it's got to be reviewed. It puts the harriers off nesting, I've seen stonechats' nests go up in flames just like that. On

hen harrier-sensitive sites it should end on 15 March, just like it does in Ireland.'

'At Bowland they're destroying the habitat,' Bill Hesketh once told me. 'Burning, burning, burning. Once the burn destroys the roots, that's it. There used to be millions of insects on the high tops for meadow pipits, young grouse and such like to feed on. Nowadays you can hit the ground with a stick. There's nothing. It's barren.'

'Some keepers are really conscientious because they know it would be counter-productive if they burnt right down to the roots,' Stephen continues, as we walk past the smouldering heather. 'They'd get no flush of green shoots to feed the grouse in a few years' time.'

Moorland management, particularly heather cutting, needs to be done sensitively and sympathetically regarding timing, and the size and orientation of the cut, as the needs of the grouse and other moorland wildlife are of paramount importance. But I remember being horrified in 2015 when I was up at Bowland to see one grouse moor criss-crossed with new roads. There will always be a landowner who wants to get his fee-paying guns quickly across the moor for another drive, one who's willing to flout the principles of moor management and scar the moorland irredeemably.

There is a striking analogy between today's devastating 'grouse-moor management', as it is optimistically called, and the effects of the 1898 *Klondike* gold rush. Initially, picks and shovels loosened river gravel to be panned for gold dust or nuggets. By 1900 most of the easy gold had

been extracted. To quote from Pierre Berton's excellent book Klondike:

> But the Klondike and its tributary valleys would be unrecognisable to the men of '98; they are choked with mountains of gravel tailings, churned up by the great dredges that for half a century have mined the creek beds. These tailings run like miniature alps for miles; and the water, changed in its course by dredging, finds its way between them in a thin trickle. The hills, still bare of trees, are marked by the hesitant lines of old ditches and broken flumes and the scars left by hydraulic nozzles.

Anything for money, and damn the environment!

In 2014 Leeds University published an incriminating report on the effects of heather burning. Researchers found that:

> the water table depth – the level below which the ground is saturated with water – is significantly deeper in areas where burning has taken place compared to unburned areas. A deeper water table means that the peat near the surface will dry out and degrade, releasing stored pollutants such as heavy metals into rivers, and carbon into the atmosphere.

I often dip into Mark Avery's blog, and recently there was one posting that caught my eye and seemed very apposite: 'I will lift up my eyes unto the hills from whence cometh my floods.' Intensive grouse-moor management dries out the hills in order to make a heather monoculture that favours red grouse at the expense of everything else – things such as black grouse, blanket bog or flood-risk

reduction. Mark emphasised that the faster precipitation runs off the hills, the quicker it gets into rivers, and that it is the rate at which the height of the peak flow builds up that causes flooding. The same amount of water spread out over a longer period of time does not present a problem.

I rang Mark and asked him about the bookshop at Hebden Bridge that was uninsured and completely flooded out. Like me he had tried to ring and offer his help, only to be told that their answer phone was full and they could not accept any more calls.

We agreed that those at risk from floods would be well advised to follow the example of the good people of Pickering in north Yorkshire, who live at the end of a steep gorge about ten miles long into which much of the water from the North York Moors drains. Pickering was flooded four times between 1999 and 2007, the last flood causing £7m of damage. After that last flood, they rejected a proposed £20m concrete wall in the centre of the town to keep the water in the river and turned instead to traditional methods of harnessing water, building dams of logs and branches that helped release rainwater more slowly, filling gullies and drains with bales of heather and planting twenty-nine hectares of woodland. Recently, when a considerable area of northern England was devastated by flooding, Pickering was able to get on with life as normal.

Stephen and I arrive at a point opposite where Beth was found in July 2012. Beyond lies lush pasture, and a sandy track runs downhill to a wooded valley with a house and some farm buildings. From there a track leads uphill to

the ravine of Thorny Grane Gill and a dry stone wall that marks the edge of the moor.

'I pointed the hand-held receiver and immediately got a signal,' recalls Stephen. 'I was going to find her, even if it took all day. I walked down to the valley bottom, where I met the head keeper. He drove me up along a track edged with dry stone walls right to the edge of the moor. And there she was, laid atop a bilberry-clad mound. Her wings were perfect. It looked to me as if she'd just flapped into that place and died.'

There may yet be a brighter future for the hen harrier. At a recent Game and Wildlife Conservation Trust seminar in November 2015 the success of medicated grit was praised for its role in staving off a predicted grouse population crash, although it was stressed that the grit should not be used continuously otherwise the strongyle worm might develop resistance to it. Equally, the number of grouse left to winter on the moors must not be allowed to get out of hand. It might indeed be time for grouse-moor owners to be more tolerant towards raptors.

A report on the population of breeding UK hen harriers was compiled in 2016 but at the time of writing was not yet ready for publication. I knew it had been a bad year for English hen harriers, with only three pairs having raised young successfully, so I ask Stephen Murphy for his thoughts. 'At Bowland, and at one other English site,' he says, 'birds returned in the spring but never took up residence and drifted off. Although there was no

disturbance at the English sites, the birds never settled down enough to breed. Perhaps the females weren't getting an adequate supply of small mammals and birds from the males – this is a really important precursor to breeding.'

Despite 2016 being a depressing breeding season, there were other hen harriers around, male birds that were not yet sexually mature, together with migrants from Scandinavia. 'I had to go up to Nidderdale in 2015,' recalls Stephen one day on the phone. 'Satellite tracking showed me that there were four hen harriers up there. On arrival I could see one female hen harrier hunting along a ridge, very low down. Now and then she'd make a grab at something. When I got up there I found several bodies of grouse that had bulgy eye syndrome, a very distressing sight. Then I saw a grouse, also infected, fluttering away from me. I ran after it and caught it. The feathers from the back of its head and neck had been ripped off by harriers, and it reinforced in my mind the fact that birds of prey, particularly hen harriers, play an important role in weeding out weak and sickly birds from overwintering stocks of red grouse.'

I remembered Stephen's plea for more volunteers to monitor roost sites so that he knows where hen harriers are and where they are coming from. Nigel Middleton and I decided to revive Roger Clarke's call for volunteers. At the end of January 2016 Nigel, Neil Chadwick and I went on a roost watch. It was a perfect day for watching, gin clear. We were in position by 3 pm and about half an hour

later the first ringtail appeared. It flew low in a northerly direction up the common, passing very close to us. It was in perfect condition, its crisp chocolate-brown plumage with a pure white splodge at the base of its tail making me certain that it was an adult female. It flew up and down before pitching in to land in the damp area.

Three other ringtails dropped in, but their arrival put up the original ringtail, and there was some flying up and down, checking everything out, before they all dropped into their usual roosting spots.

Finally, some fifteen minutes later, two silver-grey cock birds broke the skyline at the southern end of the common. As they dropped down they showed up well against the conifers. One of them was a full adult, the other a sub-adult who still had dirty brown patches on his wing coverts. Nigel called them silver gypsies, a term I'd not previously heard applied to the male hen harrier. Or had I? I had a sudden flashback to the title page of a script that Richard Mabey had written for the Hawk and Owl Trust to save the hen harrier twenty years earlier. There it was in my mind's eye, the name of the piece – The Silver Gypsy. I remembered asking Richard how he came by that name and that he told me that he had just made it up as it seemed apt. He was right, of course – it immediately evokes an image of an immaculate silver-grey male hen harrier as it floats from one hunting ground to the next.

Back in the present day, I was determined that the Hawk and Owl Trust should do something else positive to help save the hen harrier in England. Our chairman,

Philip Merricks, found a sponsor, who wished to remain anonymous, to donate a pair of satellite tags so that, through the good offices of Natural England, we could monitor two juvenile hen harriers after they fledged. Luckily, close at hand was Jemima Parry-Jones, one of our trustees and the director of the International Centre for Birds of Prey. What she didn't know about raptors wasn't worth knowing, and her calming influence was priceless.

The next problem was that there were only three breeding pairs of hen harriers in England in 2016, and there were none nesting on the grouse moors or at Bowland. I knew that the RSPB and Natural England would have first pick of the sites to be satellite-tagged, so Stephen suggested Langholm as an alternative. It was just over the border in Scotland, a mere stone's throw from England,

and any birds fledged there would be quite likely to drift south. Stephen told me that there were seven nests and that he would seek permission for us to tag two of the juveniles before they fledged. Langholm had been run down since the highly successful trials on diversionary feeding in 1998 and there were no longer any keepers there.

On 13 July Jemima Parry-Jones and Hamish Smith, a Hawk and Owl Trust volunteer, set off from Newent in Gloucestershire for Langholm, arriving there at 10.30 am to be met by Stephen Murphy and his son James. Together with representatives of the Game and Wildlife Conservation Trust and the Duke of Buccleugh's estate, they walked out through the heather to the chosen nest site. Stephen told everyone to wait while he went forward to collect the bird to be tagged. When he returned they gathered in a circle to watch as he deftly fitted the harness over the bird and sewed it up. A male hen harrier wheeled overhead, keeping an eye on the intruders.

Stephen finally set the satellite tag on a 10:48 pattern, just as he had with Bowland Beth, so it would record for ten hours and then shut down to recharge in forty-eight hours of daylight. Once this was done the juvenile hen harrier – a male – was put back into its nest. The procedure was repeated at another nest, this time on a juvenile female hen harrier. The birds were then named: Rowan for the male and Sorrel for the female.

The data received from the satellite tags was screened by Stephen and a delay imposed before it went to Hamish Smith, who converted it into an easily understandable

form, green for Rowan and yellow for Sorrel, on a day-to-day basis or as a compilation of their activities over a period. To avoid giving anything away, the information was deliberately altered if the satellite fixes showed that the birds were using a well-known roost site.

While these procedures were being refined a dedicated website was being set up so that the public could watch the birds' activities. the website went live on 16 September 2016. Both birds at first stayed close to the nests from which they had fledged, relying on their parents to bring in food. But they soon ventured further and further afield, although always returning to their natal area at dusk.

One day they saw other hen harriers making for a vole-rich foraging area and joined them. They stayed in this area for ten days or so and then cut loose. Sorrel flew north-west to a grouse moor much further north, while Rowan left Scotland altogether, speeding south to the intensively managed grouse moors around Barnard Castle in County Durham and north Yorkshire. Our chairman, in a splendid act of networking, phoned the owners of the moors to warn them of Rowan's arrival.

Meanwhile some environmental activists led by Mark Avery intensified the conflict with the Hawk and Owl Trust, rubbishing our attempts to save the hen harrier. Their main effort was directed towards raising the 100,000 signatures needed to force a debate in parliament to ban driven grouse shooting. They achieved their aim – and there was a three-hour debate – but the Conservatives were out in strength, there was no Labour support. It was

a whitewash, with nothing achieved, as no one seemed to care about the illegal killing of hen harriers. To add insult to injury Mark Avery was referred to as Mr Avery when he should correctly have been addressed as Dr Avery.

How were our satellite-tagged birds Rowan and Sorrel faring? Sorrell was well established on a grouse moor in Scotland but Rowan amazed everyone by leaving the grouse moors near Barnard Castle on 1 October and making a dash south via Cambridge to spend a few days loitering around the Kent and Essex coasts. His curiosity satisfied, he returned back to the grouse moors around Sedbergh in north Yorkshire.

Despite the well-intentioned approach of Defra's six-point action plan – published in January 2016 – to save the British hen harrier, it is throttled by bureaucracy, and the debate about the protection of this amazing raptor is going nowhere.

Stephen Redpath's joint article with Arjun Amar, 'Conflicts in the uplands: birds of prey and red grouse,' outlines the problem. 'The main impediment to success has been the unwillingness to compromise. Both sides have focused on winning rather than seeking shared solutions. Conservationists push for enforcement or management with minimal intervention, such as habitat management or feeding raptors. Grouse managers push for action, direct control, such as a quota.'

With only three pairs of hen harriers breeding in England, and none of these on a grouse moor, it is time for urgent action. I have always felt we need an

injection of new blood, which is why I think the lowland reintroduction of hen harriers from Spain, France or Russia to a suitable habitat such as Salisbury Plain or Exmoor would pay dividends. Hen harriers are great wanderers and they would soon find their way north to the Forest of Bowland, once the crucible for England's hen harrier population, and the northern heather moorlands. And perhaps Scottish Natural Heritage will soften their attitude and allow eggs to be taken from Scotland.

A major setback occurred to Defra's action plan when the RSPB recently withdrew from the working group, citing as its reason the continued intense persecution of the hen harrier. This left the Hawk and Owl Trust as the only conservation group still involved. Philip Merricks, the chairman, firmly believes that brood management is the answer to the problem and will work. He is a landowner and thus has a natural link with the owners of grouse moors.

Brood management is one of the six points of the plan put forward by Defra. The first attempt over a five-year period is funded and will run as a trial. After that it is up to the shooting fraternity as a whole to make sure that those surviving hen harriers are allowed to move around and to utilise the upland moors when they settle. The success or failure of this part of the scheme is in the hands of those with moorland responsibility. If the trial is shown to fail due to the continued persecution of the hen harrier it will cease and other less palatable methods of controlling upland game shooting will be more seriously considered.

According to Defra a trial will not be initiated until

there are two pairs of hen harriers breeding within a ten-kilometre radius. When this occurs the eggs from the second pair will be removed by experts and hatched in an incubator, with the resulting chicks being reared by experts until they are able to thermoregulate and tear food for themselves. At this point they will be transferred to a release aviary on a site chosen and monitored for prey base by Natural England, ideally as close to the home moor as possible. When the young harriers are competent fliers they will be 'soft released', with food being provided for a further period until they disperse naturally. All the birds will be fitted with satellite telemetry so that their progress can be monitored.

The Hawk and Owl Trust recently issued a statement that no members of their staff would be involved in brood management or the release of hen harriers from abroad to lowland areas in England. Both schemes would be left in the hands of experts.

At this point their survival will be up to landowners across the country, taking into consideration the normal mortality rate for first-year hen harriers.

As a precursor to the brood management trial, when there is just one hen harrier nest, this original pair will be offered diversionary feeding. This will take place during the critical six-week period when hen harriers do predate grouse chicks. From the moment the harrier's eggs hatch, day-old chicks will be placed on what is basically a hen harrier bird table near the nest in the knowledge that they will take these rather than grouse chicks from the wild. This scheme works

– trials at Langholm have shown that it can reduce grouse chick predation by 86 per cent – but it is labour intensive and not popular with many conservationists.

Diversionary feeding has been used on five occasions to rescue hen harrier nests that were in trouble. In 2006 Stephen Murphy located two hen harrier nests that were only five hundred yards apart and with only a single male bird as food provider. A consortium of bodies - the RSPB, the National Trust and National Parks - discussed the problem. Stephen suggested that the two nests should be diversionary fed, and when Andrew Heath, representing the Hawk and Owl Trust, suggested that we would need a licence to do this Stephen simply told him that as a falconer all he needed to do was get some day-old chicks and bung them in the nests. It was as straightforward as that. Job done, all ten chicks survived.

In 2007 one chick was saved from a nest on a Yorkshire moor. In 2014 two nests at the Forest of Bowland were diversionary fed – one nest (A) accepted the day-old chicks; the other nest (B) didn't. What intrigued me was that the fledging harriers from nest A raided nest B to take their day-old chicks, whom the B harriers had spurned.

James Bray, the RSPB lead at Bowland, let me know that they continued diversionary feeding in 2015 with the B pair that refused day-old chicks but they continued to ignore them.

Diversionary feeding therefore does work in practice, although I am afraid the cynic in me predicts that after the trial is over and we are back in the real world the first pair

will be wiped out before there's any chance of a second pair nesting within a ten-kilometre radius.

On 19 October Stephen Murphy let it be known that he was concerned about Rowan. He had pinpointed his location by the last satellite fix and it was now static. I was in hospital with a broken leg and I will always be grateful to Jemima Parry-Jones for letting me know the sad news rather than it coming second-hand from someone else. The next day Stephen confirmed that he had found Rowan lying face down in rushes, his tibio-tarsus leg bone shattered. He had succumbed in the same way as Beth had, and had only been able to fly a mile or so before bleeding out. A post-mortem was carried out. He had been shot.

So there remains a conflict that is unresolved. There is an arms race: sophisticated satellite tracking by the conservationists versus state-of-the-art weaponry. Some owners or managers of driven grouse moors think that they have the solution – eradicate hen harriers. But that solution is idle, cruel and unlawful.

Philip Merricks is adamant that the illegal killing of raptors must stop, but it will be difficult. The RSPB case against the gamekeeper, who on camera was seen killing a female hen harrier, was, after battling backwards and forwards, thrown out of court on 4 May 2017. We must protest ever more vigorously against the illegal killing of hen harriers.

Thankfully there are some moor owners and keepers who relish the whirling dervish, aerobatic courting display of the male hen harrier as one of the wonders of the bird world, and it makes them angry to realise that their children and their children's children may not have a chance to see a male hen harrier skydancing in England. Simon Lovell of Natural England recently told me that some grouse moor owners in Northern England have worked together to sign a Memorandum of Understanding to protect breeding hen harriers by making their moors a haven for them and other wildlife.

Could the year 2017 and those that immediately follow decide the fate of the hen harrier as a breeding bird in England?

Take heed of these words written by Professor Edward O. Wilson in his book *The Diversity of Life*:

> We should not knowingly allow any species or race to go extinct. And let us go beyond mere salvage to begin this restoration of natural environments, in order to enlarge wild populations and stanch the haemorrhaging of biological wealth. There can be no purpose more enspiriting than to begin the age of restoration, reweaving the wondrous diversity of life that still surrounds us.
>
> We must continue to vigorously protest the illegal killing of hen harriers in England. Our united voices will eventually force the government in power to listen and realise that the only way forward is to bring in a bill to license moors in England for driven grouse shooting.

Stephen recently said: 'I'm sure Beth will be remembered as a turning point. Of all the birds we've tracked, she was the one that generated the most interest and the most shock at her death. She was an exceptional bird.'

David Harsent created this requiem to Beth's memory.

Bowland Beth

That she made shapes in air

That she saw the world as pattern and light
moorland to bare mountain drawn by instinct

That she'd arrive at the corner of your eye
like the ghost of herself going silent into the wind

That the music of her slipstream was a dark flow
whisper-drone tagged to wingtips

That weather was a kind of rapture

That her only dream was of flight forgotten
moment by moment as she dreamed it

That her low drift over heather quartering home ground
might bring anyone to tears

That she would open her prey in all innocence
there being nothing of anger or sorrow in it

That her beauty was prefigured

That her skydance went for nothing
hanging fire on empty air

That her name is meaningless
your mouth empty of it mind empty of it

That the gunshot was another sound amid birdcall
a judder if you had seen it her line of flight broken

That she went miles before she bled out

ACKNOWLEDGEMENTS

Eddie Anderson and the late Mick Carroll were the driving force behind my first visit to The Forest of Bowland. In two days Stephen Murphy shuffled us between the two 'Bills' – Bill Hesketh and Bill Murphy – interspersed by long interviews with himself. Three years later Eddie and I met up with Stephen again to talk about Bowland Beth and were taken to see the location of her tragic death. I am indebted to Eddie, Mick and Steve. Without their support and knowledge this book would never have been written.

I had immense help from The Hawk and Owl Trust, particularly from our chairman, Philip Merricks, who found the funding for the satellite-tagging programme. Jemima Parry-Jones oversaw the tagging by Stephen Murphy with Hamish Smith. Later they created a Hawk and Owl Trust satellite-tracking website.

Nigel Middleton, a firm friend for many years, and Neil Chadwick revived the late Roger Clarke's monitoring of hen harrier winter-roost sites and collection of pellets for prey identification.

Terry Pickford, who has spent all his life protecting the peregrine falcons and hen harriers at the Forest of Bowland, gave me colourful information about how keepers employed by the late Duke of Westminster ruthlessly dealt with nesting hen harriers.

Steve Thornton gave an 'on the spot' account of a new hen harrier site in Wales.

Tim Birkhead offered me invaluable advice on examples of grieving birds.

The RSPB were very generous in response to my requests for information. Among those I talked to were John Armitage, Jeff Knott, Simon Wootton, Guy Shorrock, Mark Percival and James Bray.

Nick Sotherton at The Game and Wildlife Conservation Trust answered my questions about the preferred densities of grouse on driven grouse moors. Amanda Anderson of the Moorland Association was helpful whenever I rang her. Simon Lester at Langholm was always anxious to share full information when I contacted him. I'm very grateful to the Langholm Consortium and The Duke of Buccleugh's estate for allowing us to satellite-tag two of their ready-to-fledge hen harriers.

It is difficult to single out individuals who have contributed without missing out on others. So I would just like to thank everyone who has answered my calls for information.

Special thanks to Faber for allowing us to include 'A Requiem to Bowland Beth' by David Harsent; Bloomsbury for permission to include a short passage from Donald Watson's *The Hen Harrier*, the Hamlyn Group for use of an extract from *Eagles, Hawks and Falcons of the World* by Leslie Brown and Dean Amadon, H&F.G.Witherby for a passage from *The Handbook of British Birds*; HarperCollins for permission to include an excerpt from Brian Vesey

Fitzgerald's *British Game*; McClelland and Steward for an extract from Pierre Berton's *Klondike*; Cambridge University Press for a quote from *Conflict in the UK uplands: birds of prey and red grouse* by Arjun Amar and Stephen Redpath and finally to Penguin Books for a short extract from *The Diversity of Life* by Professor Edward O. Wilson.

Myles Archibald, my editor at HarperCollins, was a charming, enthusiastic collaborator. He made several first-rate suggestions which re-vitalised Bowland Beth's storyline. Julia Koppitz deftly shepherded the manuscript through the disciplines of editing and proofreading. My thanks also to Jo Walker. Her jacket for the book was simple but eye-catching. Myfanwy Vernon-Hunt's design and placing of Dan's sketches in the text was exemplary.

Dan Powell's black and white sketches have magically enhanced my descriptions of Bowland Beth's life and the world in which she lived.

Finally, I am indebted to my wife, Liza. Her enthusiasm for Stephen Murphy and the Forest of Bowland hen harriers spurred me into action and ensured that the story of Bowland Beth has been told.

INDEX

H

I

J

V

W

Y